MORE PRAISE FOR *FIND YOUR NEXT*

"Andrea has a deep technical understanding based on years of global experience in innovation combined with a rare talent for communicating important issues very simply. The approach she describes in Find Your Next *is so easy to grasp that you'll see things differently and be able to lead your teams in refreshing new directions."*

Herwig Maes
Director of Strategic Sourcing & Supplier Relationship
Management, Johnson & Johnson

*"*Find Your Next *is the missing book on every business leader's book-shelf that fits right between Michael Porter's and Malcolm Gladwell's. It's the playbook we've been wanting for hands-on innovation."*

Emily Watkins
Sr. Vice President, Innovation & Product Development
Jones Lang LaSalle

"What every business leader wants is tomorrow's news, today. Find Your Next *provides exactly that—a manifesto for innovators based on time-tested techniques. Mandatory reading."*

Tom Stat
Executive Director, Edison Universe/Adjunct Lecturer,
Farley Center for Entrepreneurship and Innovation,
McCormick School, Northwestern University/
independent innovation consultant

*"*Find Your Next *brings together a collection of insights and approaches that challenge everyone in an organization—from the CEO to the front line—to be nimble and build new muscles for rapid innovation. It disrupts the patterns of incremental growth from traditional strategic planning. The result is a process that can get your organization to market faster and leapfrog the competition."*

Alistair Goodman
CEO, Placecast

Find Your Next

USING *the* BUSINESS GENOME APPROACH *to* FIND YOUR COMPANY'S NEXT COMPETITIVE EDGE

ANDREA KATES

New York   Chicago   San Francisco   Lisbon   London
Madrid   Mexico City   Milan   New Delhi   San Juan
Seoul   Singapore   Sydney   Toronto

1 2 3 4 5 6 7 8 9 10   DOC/DOC   1 6 5 4 3 2 1

ISBN   978-0-07-177852-7
MHID   0-07-177852-7

e-ISBN   978-0-07-177854-1
e-MHID   0-07-177854-3

Business Genome® is a registered trademark of SUMA Partners, LLC, and is used herein with express permission.

Find Your Next™ is a trademark owned by SUMA Partners, LLC, and is used herein with express permission.

McGraw-Hill products are available at special quantity discounts to use as premiums and sales promotions or for use in corporate training programs. To contact a representative, please e-mail us at bulksales@mcgraw-hill.com.

This book is printed on acid-free paper.

Contents

Foreword

By Don Tapscott

Find Your Next by Andrea Kates offers a systematic approach to competing and prospering in today's hyper-competitive world. She describes how a business's core DNA, or six key genomic elements, are central to every business's success. The advice is timely, not just because the metaphor is a fresh and helpful one, but because each of these "geonomic" components is being turned on its head as the industrial age finally comes to an end.

1. The Transformation of Innovation

The pace of innovation used to be glacial. Today it has become real-time. This is being caused by a confluence of factors that center on the digital revolution.

A decade ago I likened the first iteration of the World Wide Web to a traditional newspaper. You could open its pages and read its information, but you couldn't modify or interact with it. And rarely could you communicate meaningfully with its authors, apart from sending a letter to the editor. The Web was seen as a tool to publish information that users could then browse. Today's Web is fundamentally different in both its architecture and applications. Instead of a digital newspaper, think of a shared canvas where every splash of paint contributed by one artist provides a richer tapestry for the next artist to modify or build on. Whether people are creating, sharing, or socializing, the Web is now about participating.

This is leading to a profound change in the architecture of the corporation and how we orchestrate capability to innovate. During the twentieth century, large, secretive, vertically-integrated corporate fortresses ruled the marketplace. At the time, this was the most efficient and effective business model available. In 1937 the Nobel laureate economist Ronald Coase asked: "Why do corporations exist?" After all, the marketplace was theoretically the best means to allocate scarce resources, balance supply against demand, and determine price points. So why weren't all individuals acting as separate buyers and sellers, instead of what they were doing—gathering in companies with tens of thousands of other co-workers and effectively suffocating competition within corporate boundaries?

Coase argued that the reason was transaction costs; i.e., searching the marketplace for the right product and negotiating its purchase. Companies found that it was more cost-effective to perform as many functions as possible in-house. But suddenly, this is no longer the case. Many behemoths are losing market share to more lithe competitors. Digital technologies slash transaction and collaboration costs. Smart companies are making their boundaries more porous, using the Internet to harness knowledge, resources, and capabilities outside those boundaries. They set a context for innovation and then invite their customers, partners, and other third parties to co-create their products and services. In most industries, companies innovate and perform better by creating networks or what I call "business webs."

The mantra "focus on what you do best and partner to do the rest" is serving most leaders of the global economy well. In the past a company would outsource functions and ask for weekly or monthly status reports. Today the status reports are 24/7 as companies integrate their networks. Rather than offloading a process, companies now collaborate.

2. The Transformation of Marketing into a World Where Customers Can Be Inside

Just as it makes sense to think of talent being inside *and* outside your organization, it makes sense to view customers in the same way. Customers are no longer external entities that your company pushed products and services to. They have become part of your business network. Rather than just focusing on them you can engage them. Rather than creating goods for them, you can co-innovate with them. Rather than having customer centricity, you can build customer communities or, as Kates calls it, "customership through dynamic conversation."

Customers also have unprecedented knowledge and an ability to scrutinize companies, share information with others, and organize collective responses. This gives them great power, turning most of what we know about marketing on its head.

With the Internet becoming ubiquitous, the Four P's—product, price, place and promotion—no longer work. The paradigm was one of control, simple, and unidirectional: firms market to customers. *We* create products and define their features and benefits; set prices; select places to sell products and services; and promote aggressively through advertising, public relations, direct mail and other in-your-face programs. We control the message.

The Internet transforms all that. Products are now mass customized, service intensive and infused with the knowledge and the individual tastes of customers. Because companies must constantly innovate, product life cycles collapse. Enabled by online marketplaces, dynamic markets and dynamic pricing increasingly challenge vendor-fixed pricing. Online communities upend control. The brand is also becoming more complex construct—one owned by the customer.

3. The Transformation of Talent, Culture, and Leadership

Too many businesses are still stuck in the old unproductive command and control hierarchy, which divides the world into governors and the governed. The middle managers in between acted, as business thinker Peter Drucker put it, as "relays—human boosters for the faint, unfocused signals that pass for information in the traditional, pre-information organization." This was a hierarchy and your job was to move up, and fulfill the goals determined by your boss—or his boss, or his boss's boss. Innovation, creativity, giving customers better service, or creating products were typically not part of the picture. You hung in with the company until you retired or were fired. You were the "Organization Man." This model obviously fails in an economy driven by innovation, knowledge, immediacy, and working via the Internet.

Probably the biggest factor challenging conventional management is the massive generation of young people entering the workforce. I call them the Net Generation, and they outnumber their baby boomer parents. Younger workers bring with them a new ethic of openness, participation, and interactivity, and cannot imagine a life without Google or Facebook always within reach.

Unlike their parents who watched 24 hours of television per week when they were young, today's youth are grew up interacting. They go online to scrutinize, authenticate, collaborate, and organize. They can't imagine a life without the tools to constantly think critically, exchange, challenge, authenticate, verify, or debunk other people's views.

As employees, they are motivated by non-traditional factors. They are willing to work hard but also insist on work-life balance. They yearn for constant feedback rather than an annual performance review. They don't respond well to traditional forms of supervision or even hierarchies, preferring to work in collaborative teams. And my research suggests that companies that embrace them and their culture tend to perform better.

Leadership is changing, too, as the old command and control models are being replaced by models that harness collective intelligence. Peter Senge was right 20 years ago when he admonished that the person at the top can't learn for the organization as a whole any more. But his ideas for creating organizations that can learn have come of age.

4. The Transformation of Process by Collaboration

The rise of unstructured work, collaboration, a new generation of knowledge workers, and new suites of collaborative tools are also changing the business process. Today, many business processes are hangovers from the industrial age of mass production. My colleague Tammy Erickson explains what they were focused on: products—how to develop, manufacture and sell goods; scale—how to produce "enough" to meet expanding consumer demand; scope—how to distribute for a mass consulter market; and quality and cost—how to standard and maximize efficiency. Organizations that mastered these capabilities dominated the twentieth-century economy. They benefited from bureaucratic systems that optimized core processes and a clear division of responsibility that needed only top leaders to worry about the overall goals.

Today the challenge is to orchestrate intelligence. Companies must use one's particular knowledge and capacities in ways that contribute to the success of the whole. They must combine different types of knowledge and expertise to come up with something better, harnessing the smallest units of knowledge and continually improving processes and routines. And they must customize relationships with customers, suppliers and others, detecting and responding to market and environmental shifts.

We are shifting from closed and hierarchical workplaces with rigid employment relationships to self-organized, distributed and collaborative human capital networks that draw knowledge and resources from inside and outside the firm. We have seen swift adoption of new tools, such as wikis, blogs, tags, collaborative filtering, digital brainstorms, telepresence, RSS feeds, and more. I call it the wiki-workplace. Work has become more complex and team-based. Effective employees have more social skills, are smarter at using technology, and are more mobile. The result is much faster innovation, with the added benefits of greater agility, reduced costs and improved responsiveness to customers.

5. The Transformation—and Spread—of the Secret Sauce

In the past, companies sought traditional sources of competitive advantage—and rightly so. As Michael Porter explained years ago, competitive differentiation was the foundation of growth, shareholder value, and long-term success. Some writers like Michael Tracey lecture managers that they must focus on one competence, be it product superiority, customer intimacy, or operational excellence.

But the demand pull of the fast competitive global business world and the push from the digital revolution is requiring companies to find that special recombination of their corporate genome to achieve any kind of success not simply ephemeral. Kates comes at this from the customer perspective and rightly so, for that's where differentiation and competitive advantage is tested, evidenced, and realized. In these pages you'll learn how to "Find Your Secret Sauce."

She's also right to emphasize today's hyper-transparent world when cooking up your sauce. Today customers can use the Internet to help evaluate the true worth of products and services. Employees share formerly held secret information about corporate strategy and management. To collaborate effectively, companies share intimate knowledge with one another. And, in a world of instant communications, whistle-blowers, inquisitive media, and Google, citizens and communities routinely put firms under the microscope. So, if a corporation is going to be naked—and it really has no choice in the matter—it had better be buff.

Smart companies that embrace transparency will prosper. They recognize that proactive transparency increases corporate success and will be an increasingly important source of differentiation in the future. But which corporate functions need to be involved in managing transparency to ensure success, and what are the unique leadership responsibilities of chief experience officers? What are the key classes of information, and what should be open or closed? Which stakeholders should get access to information, and how often? What social media networks should companies monitor, and how can they engage? And what is that special combination that can make a brand stand out and really pop in the relationship marketplace?

6. The Emergence of Trendability

I've never liked being called a "futurist" because I'm of the school that "the future is not something to be predicted, it's something to be achieved." My view is not just a philosophical one—it's increasingly a matter of practicality. Let's face the facts: it's getting tough to predict just about anything.

Kates agrees, encouraging readers to stay ahead of the unpredictable. With sophisticated business analytics, early warning systems, and real-time response, it is getting easier to identify trends and to not only respond to them but to achieve and shape them. As Kates points out, early sighting of trends can lead to market dominance. Abandon the notion that you can predict what comes next, and focus instead on being able to "adapt, refocus, and thrive in the new competitive landscape." She explains why responsiveness to change is the "new killer application in the corporate arsenal."

Too many companies see only what is in plain view, with initiatives that are often based on stale plans and projections that are no longer realistic. The result is that companies can head in the wrong direction. But as Kates notes, "Technological innovation can threaten the touch of personal service. But it can also create opportunity." A great example: how the iPad is changing retail. Hairstylists can offer clients previews of different cuts, while fast food restaurants can show customers exactly what they are getting when building their perfect burger.

To see the trends that may shape your industry tomorrow, look at what is shaping other industries today. Step back from the day-to-day and reflect on early signs of change. Kates focuses on three core categories: 1) technology, 2) societal shifts and economic change, and 3) customer experience. The goal is to create an informal dashboard to map trends that could translate into competitive threats.

Find Your Next

When you add all this up, ours is indeed a time of great change and arguably much of what we know about management today is becoming obviated. That is why Find Your Next is such a powerful weapon needed in every manager's arsenal. It embraces the shift from industrial models to models for the twenty-first century. And it's a great read—packed with insightful stories and tons of practical advice.

Read, enjoy, and prosper.

Don Tapscott is the author of fourteen books, including Gro*wn Up Digital: How the Net Generation is Changing Your World* and, most recently (with Anthony D. Williams), *Macrowikinomics: Rebooting Business and the World*. He is an Adjunct Professor at the Rotman School of Management, University of Toronto. Twitter: @dtapscott

Acknowledgments

Stay married to a medical scientist—as I've been—long enough and he's bound to influence your thinking. Engage in conversation with a pioneering biologist like E.O. Wilson, and you're likely to be propelled to look differently at the world of business innovation.

Flanked by my husband's perspective on DNA and genomes on one side and Wilson's encouragement and inspiration on the other, I began my journey to discover what I call the "business genome" approach. I engaged in discussions with a diverse array of business leaders who shared insights about how they changed the games for their companies. My goal was to understand the secret sauce of business innovation—how an organization truly finds its "next."

A few years ago, I sat with E.O. Wilson at breakfast the morning he was to help spawn the *Encyclopedia of Life* (an online biological encyclopedia) with support from the TED (Technology-Entertainment-Design) Prize. Just before he stepped up to the podium, he asked me what my vision might be for a project of that scale. I explained the model I envisioned for the business genome process, a method I thought would unlock the potential of business growth, and briefly outlined the 15 years of observations that led me to uncover what really distinguished the winning strategies from the stagnating ones. We talked about the power of taxonomy (he called it a "stamp collection" mindset) when combined with metaphorical thinking—the ability to see similarities in companies across industries—and I told him I was interested in combining those two outlooks to create a new discipline for business innovation.

Shortly after that breakfast, I continued to pursue the concept of developing a Business Genome process and met with a broad cross-section of business innovators: Tim Westergren from Pandora; Neil Hunt from Netflix; Bill Bragin from TED; Don Tapscott, author of *Wikinomics*; James Fowler, author of *Connected*; Thomas Goetz from *Wired* magazine; Jack Myers from Media Advisory Group; Peter Diamondis from the X Prize Foundation; Craig Donato from oodle, Alan Jacobson from ex;it; Cooper Bates from the LA Idea Project; Doug Solomon from IDEO; Bill Villarreal from Dolby; John Danner from University of California at Berkeley's Lester Center for Entrepreneurship and Innovation; Clay Phillips of General Motors; Cameron Sinclair, founder of Architecture for Humanity; Mark Bouzek from BP, Darryl Drenon from Cisco Systems; Cameron Blakely from Sysco Foods; Douglas Newby of Architecturally Significant Homes; journalist and author Catherine Crier; and dozens of other individuals who would provide important perspectives to the development of the Business Genome project. Every member of that group validated my initial concept, which was really just a vision at first, and what I soon realized was a business frontier.

The group's consensus was clear: define the Business Genome by finding stories that would illustrate a new way of thinking for business leaders, based not on traditional MBA skills, but on cross-industry pattern recognition. Catherine Crier, in turn, introduced me to Jan Miller, Shannon Miser-Marven, and Lacy Lalene Lynch—the team from literary agency Dupree Miller & Associates, who said, "The next step is to write a book." They came on board to represent me and that's how *Find Your Next* came to be.

I'd like to acknowledge the insights of clients and colleagues like Nick Pudar, vice president of planning and development for OnStar/GM; Lane Cardwell, president of P.F. Chang's; Bruce Northcutt, president of Copano Energy; and, Matt Winter, president and CEO of Allstate Financial, for allowing our Business Genome team to walk in their shoes for many months, refining our tools and methods along the way.

The 15 years and 250 strategy projects that fueled the initial thinking behind the Business Genome also included many other sources of inspiration, from Royal Dutch Shell to Hyatt Hotels to entrepreneurs throughout the country, all of whom provided the backdrop for the lessons shared in *Find Your Next* (many of which had to be translated into generic versions to maintain confidentiality).

To Mary Glenn, Gary Krebs, Sara Hendricksen, Ann Pryor, Gaya Vinay, and their colleagues at McGraw-Hill, Richard Rothschild and David Andrews of Print Matters, thank you for helping plug these ideas into a larger ecosystem of thought, so that they would have a chance to influence other writers, leaders, innovators, and communicators.

And to each of the individuals who generously shared their realities and the story-behind-the-story for every refreshingly-told narrative, thank you:

Chris Bradshaw of Autodesk—immersion in a new consumer mindset to discover the newest waves in design

Rebecca Campbell of Chase Card Services—a new perspective on corporate change

Sharon Richmond of Cisco Systems—insights into a culture of innovation

Tim McEnery of Cooper's Hawk Winery & Restaurant—the initial "hunch" that led to a multi-unit restaurant

William Teuber of EMC Corporation—the ups and downs of a highly competitive landscape

Mark Vachon of GE Ecomagination—a down-to-earth telling of a multi-billion dollar innovation story

Vijay Iyer of GM/OnStar—tracing the 'aha moments' from the technology capabilities to the consumer preferences

Steve Elefant of Heartland Payment Systems—lessons about innovation driven by a rising tide of risk

Jamey Rootes of the Houston Texans—cross-industry insights inspired by the likes of the McDonald's drive-through

Tom Smith, David Nadelman, and Matt Adams of Hyatt Hotels and Resorts—analysis of a new brand that extends the guest experience

Danae Ringelmann of IndieGoGo—the path from Wall Street analyst to entrepreneurial pioneer

Herwig Maes of Johnson & Johnson—the logistical lessons learned from Nike and Skil

Emily Watkins of Jones Lang LaSalle—innovation tools that spread ideas

Don Spetner of Korn/Ferry International—the honest account of when it's better to take a measured approach to innovation

Scott Wilson of MINIMAL—the force behind the $900,000 miracle of the NanoWatch/Kickstarter campaign

Rick Federico, Lane Cardwell, and Mike Welborn of P.F. Chang's China Bistro—the lessons on rethinking a beloved brand

Alistair Goodman of Placecast—insights on when new technology isn't the only platform

Adam Kanner of ScoreBig—inspiration on building an entrepreneurial culture that supports innovation

Dan Gross of Sharp HealthCare—the organization lessons that led to winning the Baldrige Award and reshaping the healthcare experience in the process

John Winsor of Victors & Spoils—shaking up the models of traditional advertising

Brad Inman of Vook—looking at publishing through a new lens

Thanks also to colleagues who shared their thoughts and reviewed the concepts within: Clyde Beahm, Jonathan Berman, Larry Burns, Nelly Colapinto, Dan Demeter, Susan d'Herbes, Susanna Finnell, Ronnie Hagerty, Amy K. Hutchins, Kathryn Jaroneski, Jodie L. Jiles, Ben Kern, Daniel Kraft, Mary Lawton, Maria Lefas, Suzy Lebovitz, Elizabeth Luff, Dan Mapes, Nan Morris, Cathy Nunnally, Scott Parazynski, Tim Schaaf, Dave Schotterbeck, Lisa Tran, Bill Villarreal, Cindy White, and Peter Write—and the team that tested the genomic concepts in CoLabs, our hands-on workshops and strategy sessions, which included Jeffrey Frey, Dennis Irvine, David Aronica, Susan Bryan, David Goldstein, and Thomas Stat.

In the spirit of cross-industry innovation, I hold special appreciation for the contribution of Perry Farrell (of Jane's Addiction and Lollapalooza) for linking the lessons of *Find Your Next* to the world of live entertainment, and to Lacy Lalene Lynch, Al Hassas, Patrick Pharris, and

Daniel Decker, who were brave enough to take an unconventional, cross-industry approach to building a readership and extended community for *Find Your Next*.

For helping to think through the initial algorithms behind the Business Genome engine and the viral plan for building a repository of business successes: close colleague Jeffrey Frey, Tim Westergren of Pandora, Nick Pudar of OnStar, Clay Phillips of GM, Neal Hunt of Netflix, John Danner from University of California at Berkeley, Jed Sundwall, and James Fowler, author of *Connected*.

A special acknowledgment to Don Tapscott, best-selling coauthor of *Macrowikinomics*, whose thoughtful foreword framed the book within a broad, global context. And thank you to Seth Godin, author of *We Are All Weird*, whose cover endorsement speaks volumes.

This book would not have been possible without the contributions of Emma Kate Tsai, who read every word and helped to hone my voice "bird by bird," and Jeffrey McKay, the illustrator who shaped those words into visuals that enhanced my message.

And, finally, to Phil, Rhoda B., Rhoda S., Surelle, and Milt—my heritage—and to Russ, Adam, and Zach—a cross-industry family that inspired this project from its very inception.

Introduction

Follow That Hunch

From the age of 11, Tim McEnery worked his way up the food chain of food, first washing dishes in a local café, then waiting on every kind of diner in a variety of restaurants, and eventually managing an eating establishment owned by Aramark. He had never done anything else, and he didn't want to. Food was in his blood. He'd invested the thousands of hours it took to know every perspective of the restaurant business intimately, from the back-of-the-house—washing and drying—to front and center—waiting on tables and understanding customers—to, now, the bird's eye view—management.

McEnery's head was no longer so focused on serving a plate of food, or cleaning one that had just been cleared. Instead, his mind had become full of ratios of food costs and facts about failure rates, and his heart swept up in a dream-come-true inspired by a first date with his soon-to-be wife. They had just enjoyed a wine tasting at a very unique spot just outside Chicago—where McEnery ended up working a couple of years later. He was still working at an Aramark-run restaurant as his main job, and from time to time had fleeting thoughts of running his own place one day. Sitting across from the woman of his dreams on what would become a historic night, his daydreams came into full focus. He wanted winery experience, and suggested to the winery—as a way of getting in the door—a "wine maker dinner" that his restaurant would pay for. All the winery's people had to do was show up with the wines. And so, they did.

It was really nothing more than a crazy *hunch*. And began as a way of getting the experience he needed to later start something of his own. He knew that running a restaurant was one of the toughest ways to make a living and that starting one could be a fast path to losing your shirt if you weren't careful. But McEnery couldn't get the idea out of his mind; he kept wondering if he could possibly recreate the magic of that special dinner event with an entirely new format. It was risky and challenging, and revolutionary. Besides, he'd never run, or started, his own business before. Still, he was ready.

But, could he do it?

What McEnery had in mind was a wine-infused eatery, and he would have to pave the path from chain restaurants and boutique cafés to something completely different. And, eventually, he would have to build a community of wine enthusiasts who would become cult fans of his concept. He found himself at a crossroads. What McEnery saw in his mind didn't exist, not yet. Because he wasn't just imagining a restaurant that made its own wine. McEnery's vision was larger than that. What he envisioned wasn't just another eating establishment that would add to the already growing and segmented genre, but something that would dramatically shift the dining-out model, and he wondered if the new dynamics of community building through wine clubs couldn't also be added to the mix. And by pursuing this new concept, he was proposing to write an entirely new chapter in restaurateuring, in the flavor of date-night restaurants like P.F. Chang's and the Cheesecake Factory. It would mean transitioning away from candlelight and white linen, and it might also require new rules, new venues, new décor, new expertise, new technologies, and new job descriptions. Perhaps he wasn't just proposing to add pages to the treatise on dining, but in fact penning a brand new book.

Danae Ringelmann was sitting on Wall Street as an analyst, and she was comfortable there. But she didn't spend all her days and nights crunching numbers. In her spare time, she volunteered in the world of theatre, linking playwrights with producers and money. That's when she began to notice a financial disconnect between artistic creation and artistic funding: theatre arts were paid for either by the bureaucracy of government support or the unpredictable donations of wealthy individuals. In other words, politicians and rich people ultimately decided which plays ran and which plays never got off the ground. No other groups were represented in the backstage of money. The theatre-goers were a different story. The audience for independently-produced plays and films—passionate people who really cared—consisted of a much more diverse group with a much more varied strata of income. Their dedication would have them waiting hours to meet the playwrights and screenwriters, spending precious time they didn't have learning about the process of invention, and following the project from inception to staging to blocking to opening night. They were more than just fans of the plays, films, and productions they entrenched themselves in and as a result, they did so much more than simply buy tickets and show up to fill seats. They were actual zealots when it came to supporting the voices of the inventors of the oeuvres, and they could prove it.

This disparity meant Ringelmann was smack dab in a second, real-life version of "the long tail," a phenomenon conceived by Chris Anderson in 2006 (Anderson, 2006) that originated when independently produced music began getting support directly from its listeners. No longer was it a requirement for a major producer to cut a CD; independent music groups could now fund the production of new songs solely from their fan base. And that same thing could happen for performing arts. Ringelmann had a *hunch* that she could turn theatre upside down by changing the way it fed itself. In pursuit of that idea, she enrolled in business school to find out if she could enact such a world-altering shift in the way things had always been done.

Nick Pudar had been with GM's OnStar division—a department that created a factory-installed safety, security, and navigation system in GM cars and trucks—off and on since its inception. Through his tenure there, Pudar watched as the customer's relationship with technology changed over time, seeing much greater possibility for OnStar than simple auto-directional advice. People were beginning to relate to personal mobile devices and technology in new and interesting ways, and OnStar subscribers were no exception. They were no longer just "using" the service, but depending on it, trusting it, connecting with it, and committing to the human voice that led them through dark roads at night, "held their hands" in the aftermath of a car accident or vehicular emergency, and could help them in case they locked themselves out. OnStar became the driver's access to information and resources in every situation. And Pudar, a master strategist (and a deadly serious amateur magician), had an epiphany. What would it take to bring the solace of OnStar to everyone who needed connectivity in any capacity? Not just drivers of a GM car? Pudar saw his product—24/7/365 service—and the brand as having much farther-reaching potential. Because it wasn't just the community of GM that needed OnStar's connectivity, but anyone who got into a car and maybe even outside of it. Still, in a company the size of GM, it would take more than a cool idea to grab the attention of management, and a lot more than a *hunch* to capture the loyalty of the public. Pudar was inspired. But where would Pudar construct the path that would get him from inspiration to execution?

What McEnery, Ringelmann, and Pudar had in common was passion, inspiration, and desire. And a *hunch* that things could go beyond where things had always been. Theirs was a feeling everyone in business could identify with: that experience, industry knowledge, and observations of emerging trends could intersect to create a business opportunity that transcended today's boundaries. And they faced a challenge everyone in business could identify with, too: that the tools and skills they'd learned and used weren't sufficient for moving them toward what had never been done before. Their toolkit was tailored to what *was*, not what could be.

McEnery, Ringelmann, and Pudar first tried all of the old methods—like the traditional SWOT analyses, forecasts, benchmarks, and predictive models—on their new ideas but found each method fell short. Simply improving on those past processes and traditional financial analysis measures couldn't create future opportunities, either. Innovation tools informed their creative concepts, but they weren't enough to flesh out the chart of sustainable business growth. They needed something to enable them to see ahead to the future sooner and something that would translate their ideas into new models that could gain traction and thrive.

More specifically, McEnery couldn't take the spreadsheets from existing restaurants and merge them with other potential "non-food" revenue sources to create a financial pro forma of what might come next. How could he anticipate the future potential for his concept by looking at financial projections that applied to dining-only establishments?

And likewise, Ringelmann couldn't apply her Wall Street analyst skills to a world of commerce that had only recently become a possibility. It was a new, more democratic platform, one in which funding for artistic works could marry the freshly-minted dynamic of social media to allow for input and participation from the supporter of creative works.

Pudar, too, couldn't adhere strictly to an analysis of current strengths, weaknesses, opportunities, and threats to fully capture the power of what he and his team envisioned. So instead, Pudar and his colleagues would have to develop their own process to take OnStar from its longstanding version—a manufacturer-installed, in-car safety and security system—to something much broader in impact and potential—a universal device that could be purchased at any electronics store and installed in any vehicle. And so they did. At the 2011 Consumer Electronics Show in Las Vegas, Pudar and the OnStar team introduced OnStar FMV (For My Vehicle). A remote rear-view mirror unit—compatible with more than 90 million cars on U.S. roads, available at retail stores nationwide, and offering OnStar's award-winning safety, security, and connectivity. This new OnStar product fulfills the promise of connecting drivers and their vehicles to a wide range of services. FMV is an example of the opportunity that can arise from a new way of thinking.

All this boils down to one thing for sure, that adding up the elements of the past doesn't get you to the full impact you can have in the future. But what does? Something else, something different, something new, something flexible. McEnery, Pudar, and Ringelmann had to start over, they had to create new tools and processes, and they did, pioneering brand-new, tailored approaches that worked.

A New Playbook, a Fresh Look

The new way of doing business that all three came to find is what this book is about. *Find Your Next* illustrates the new principles today's leading organizations must follow to survive in today's competitive business environment. Their real stories, along with hundreds of experiences I've had and dozens of interviews I've held in my career as a business strategist and consultant, demonstrate a new way to respond to the emerging realities of business growth that all of us, as business leaders, now face. What each and every story or case study has in common is a set of simple truths—business breakthroughs that combine the instinctive hunches of what might be with revelatory insights into cross-industry trends in plain sight. With that new philosophy to guide them, companies can now thrive and teach the rest of us how. They've created a powerful new dynamic for growth that hasn't been systemized. Until now and *Find Your Next*.

Find Your Next is a new playbook with new rules, based on the trade secrets that successful business leaders like McEnery, Ringelmann, and Pudar discovered to their great success. It's a book for the millions of business leaders who want to turn their hunches about what *might be* into what *can be* and *what is*. It offers a new perspective that integrates today's realities—a faster Web 2.0 Internet world, a transparent, global marketplace, and a post-*Mad Men* dynamic—and new tools to respond to them. In this see-through universe of *commerce*, customers, consumers, clients, colleagues, suppliers, and competitors have all turned up the volume on their voices and challenged us to listen.

Find Your Next demystifies the magic behind blockbusters like McEnery's Cooper's Hawk, *Restaurant News' 2010 Hot Concept*, Pudar's OnStar, *Edmunds.com top ten car technology trend* last year, and Ringelmann's IndieGoGo, recently marking her as one of *Fast Company*'s most influential women in technology, not to mention hundreds of other organizations like Jiffy Lube, P.F. Chang's, ScoreBig, Vook, Allstate Financial,

JP Morgan Chase, Victors & Spoils, CareFusion, Hyatt Hotels and Resorts, Sharp HealthCare, G.E.ecomagination, and EMC Corporation.

The Structure That Organizes the Chaos

This book taps into new patterns that have been proven to drive business growth, founded on the elements of the "genomic" approach. In genomics, scientists are able to identify, map, and learn from patterns of an organism's DNA. In business, we can do the same thing by breaking down the core "DNA" of a company into several basic elements. With that insight, we can see how all our companies are the same, and what patterns they fall into, regardless of industry.

The core DNA, or six key genomic elements, are central to every business's success. By focusing on them, and looking for connections from industry to industry, we can breathe a little life into our companies and our bottom line.

The Six Elements of the Business Genome

With the business genome classifications, companies can organize their "dashboards" around categories, or core DNA, that reveals new opportunities for growth:

1. Product and service innovation—the invention of offerings that resonate.
2. Customer impact—a sustainable community of support.
3. Process design—alignment of the "how" of a business with the evolving "what" that customers need.
4. Talent and leadership—the culture that will move a business forward.
5. Secret sauce—the recipe of differentiation and competitive advantage in a new world of unprecedented transparency.
6. Trendability—the foresight to see the future more quickly and adapt more rapidly to shifts in the landscape.

These six elements can function as a strategic lever to move a company from a stagnating today to a more dynamic future. By focusing on the basic elements of a company's "genome" (a concept we'll explain in greater detail later in the book), companies can get in touch with evolving customer preferences and opportunities on the horizon that they can learn to respond to more nimbly. And that means mapping your company's genome against innovations of proven success, a process translated into four simple steps in this book.

The *Find Your Next* Steps: A Process That Will Catch You Up with the Speed of Change

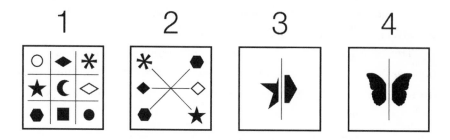

The Find Your Next approach is a process of four steps, based on a new perspective of a future that is foreseeable by scanning innovation across industries and adding them together to create inventive possibilities. The steps combine the art of instinct with the science of experience.

1. **Sort** through the options for your company and assess your hunches.
2. **Match** your genome to successful businesses that have already steered themselves in the direction you want to explore.
3. **Hybridize** your company by grafting the ideas that work in other companies onto your own. Don't be afraid to venture into the unknown.
4. **Adapt and thrive** by breaking out of old habits and fostering new traditions in your business that will enable you to take advantage of a rapidly-evolving business environment.

Crossroads of a New Infrastructure

Crossroads are everywhere. Sometimes they result from a crisis, when the world changes overnight. (Like when Napster shocked the music industry, illustrating to record companies their vulnerability in the new age of online digital music distribution[Levy, 2000]. As a result of that revolutionary crisis, the term "Napster moment" has been added to our business vocabularies, and describes a phenomenon the publishing world is now in the center of.) Other crossroads can emerge outside of a calamitous event, when we find ourselves in a shifting industry. (Think, for a second, about food, and when grocery stores started offering more and better prepared foods, and customers made the gradual cross-fade from eating in restaurants to taking chef-made meals home.) And, still other times, crossroads simply appear as a gut feeling to a pioneer with a vision.

This book is for any business leaders who somehow ended up at a crossroads for whatever reason. *Find Your Next* is the parable of every fork in the road and the inherent lesson within. It's the story of commerce and customers and leaders and culture. The characters are universal and local, the venues global and domestic, and the wares standard and high-tech.

What might seem like an abrupt urgency for a new view of business isn't what it seems. It has actually been sneaking into boardrooms and break rooms in virtually every corporation, onto the whiteboards of every start-up, and into the dreams and imaginations of every aspiring entrepreneur for at least a decade. This set of realities isn't new, per se, but it *is* the first time someone's writing it down.

The point is this: The ability to see the future more quickly and respond more nimbly is the new "killer app," *the* skill that will enable some companies to thrive and force others to go extinct.

Find Your Next doesn't dwell on the path to extinction, through recent high-profile bankruptcies that have shaken the world. It doesn't focus on the companies that couldn't predict the future or the industries that were here yesterday and gone today.

Instead of focusing on those stories, it teaches from them. It sheds light on dinosaurs by telling the actualities that made them prehistoric. It explains how every past relic acts as a symbol of a powerful and universal lesson learned. It has woven such insights into new business fundamentals and tangible steps that we can all embrace to avoid uncertain death and catch up to the future today.

No Longer "Business as Usual"

The new realities of speed, transparency, global reach, and customer dynamics declare that business can never be "as usual" again. As you'll see from the stories that follow, companies like Allstate, Cisco Systems, JP Morgan Chase, Jiffy Lube, and P.F. Chang's have responded to a new climate that demands we move beyond our original business models. Relatively recent entrants to the business landscape like Score Big, Vook, IndieGoGo, Victors & Spoils, Cooper's Hawk, and MINIMAL were able to launch as "greenfield" start-ups, using innovative growth strategies that capitalize on new competitive realities. In every case, the leaders who transformed what they saw as potential into actual success had to make up new rules and abandon traditional playbooks. The lessons learned by companies just like theirs have informed the *Find Your Next* process this book is based on.

Over the past decade, virtually every company in every industry has taken dramatic new tactics to move the ball forward. But, until now, the elements of their systems haven't been documented and translated into a unified process that every business can follow. It is exactly in that spirit—of documenting what really works in today's new world—that this book has been conceived.

Part
1

THE BUSINESS GENOME APPROACH

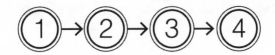

Chapter

1

The Business Genome: The Key to Next

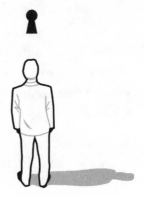

Before anything else happens, an idea is born.

Democratize the automobile. Naturalize cleaning. Give shoes to the needy. Smooth out ticket pricing. Crowd source creativity. Communalize coupons. Shelve books on cell phones. Deliver the world overnight. Bring Napa Valley to wherever you are.

Then, a business emerges. Many do: Ford, Method, Toms Shoes, ScoreBig, IndieGoGo, Groupon, Coca Cola, Vook, FedEx, Cooper's Hawk.

Sometimes an organization—a brand—develops organically, inspiring a group of people to naturally flock to it and engage with it with little persuasion necessary. Other companies show up ahead of the crowd that will eventually come to support it. In either case, as the idea evolves, so do the processes, systems, people, products, and modes of distribution that will sustain it. And the company carves out its niche in the competitive landscape.

Corporate growth continues down a path, creating silos of expertise in finance, accounting, product development, research, marketing, and talent, until the organization defines its point of difference in a mushrooming world of customers.

Then, one day, someone in the organization will have an idea, a hunch.

- *"I wonder if we could take this concept national."*
- *"I think we could expand our product line."*
- *"What if we could solve a bigger problem for our customers?"*
- *"I bet we could tap into the economic trends much more powerfully."*
- *"We're missing a big opportunity."*

There the process begins: writing the company's next chapter, uncovering its next set of opportunities, realizing its next competitive edge, developing its next area for reinvention, locating its next group of customers. Or, for an entrepreneur, planting the seed of a brand new company altogether.

And, until now, that journey would have started with dissection. The company would measure its current strengths, weigh its component parts, evaluate its past results, and scrutinize its industry peers in search of its "next." The silos of capability that grew and developed to support complex business requirements now come to define the options for analysis. The range of potential places of where to go next may look a lot like what a company has already done, but instead is actually bigger, better, or different.

Analyzing humor is said to be a little like dissecting a frog: why bother when it only ends up killing the frog. In other words, humor loses its power when pulled apart. Likewise, a company can't be understood by simply looking at each of its components in a vacuum. The problem with dissection and analysis of parts of a whole is what it doesn't reveal. It can only show the obvious in any one of the individual component parts. But there is more to be seen in the bigger picture.

The strength of the most powerful opportunities is born from the cohesion of disparate elements from multiple strategies, and an incorporation of trends that can drive business in the future. The best way to write the next chapter for your organization is not to tweak each structural

ScoreBig: An Industry-wide Solution

Adam Kanner had an idea. He envisioned a world in which fans could save substantial money on tickets to sporting events, concerts, and shows, and entertainment and sports venues could solve the problem of 40 percent of tickets going unsold. But he needed to ensure that these tickets could be priced at a discount without damaging the brand of the seller, or cannibalizing those tickets sold at full-price. To get there, he abandoned the SWOT analysis lessons from his MBA courses and didn't just create incremental enhancements to ticket pricing. He had worked for the NBA; he knew all about that. Everyone did. But he also knew that reengineering existing ticket pricing wouldn't go far enough. The gap between what the public needed—an entirely new rulebook for sports and entertainment ticketing—and what the sports franchises and entertainment promoters were offering at the time—a tiered pricing structure—was too large to address with piecemeal solutions. Instead, Adam envisioned the new solution and set out to get there, constructing a powerful way to achieve it that cobbled together elements from e-commerce, social commerce, dynamic ticket pricing, direct marketing, sports, and entertainment—and gave birth to ScoreBig.

element individually but to create something bigger and all-inclusive—and then a game plan for how to get there.

1 + 1 + 1 = 5: The Value of a Systems Approach to Change

Matt Winter, CEO and President of Allstate Financial, believes in the value of what he describes as the systems approach to the sale of financial services: a multifaceted, interdisciplinary process with which the contribution of any individual component—product, for example—is enhanced by the impact of the other parts of the overall capacity—price and people, for example. As Matt Winter sought to reinvent his company, he realized it was time to recalculate the entire equation of consumer satisfaction. He thought it through, eliminating the obvious: simplification of the policy issue process, offering the same products being sold online, or updating slogans and branding. That approach would mitigate the true power of the change he envisioned. Winter knew that simply changing individual components wouldn't get Allstate Financial to its next level. The power of the company's reinvention, Winter believed, one that led to the development of an entirely new product set, customer experience, and distribution model and economic framework, had to come from the power of the combined, synergistic effect of multiple components, and not the individual effect of alterations to any single element.

Good For Life: Thinking Like a System

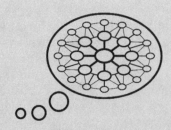

Good For Life represents more than a new product for Allstate Financial's company, customers, and distribution ecosystem. It combines simultaneous change in process, product, distribution, compensation, and brand, and establishes the capability for all of those forces to feed off one another. Winter decided to depart from the way the organization had been approaching change prior to Good For Life—isolated product, system, and brand changes—and move toward a more system-wide approach. He likened the dynamic to the world of drug interaction, in which one drug's efficacy is greatly enhanced by the presence of other drugs administered at the same time. "To generate massive change in a system like Allstate Financial's, which is what we were trying to do, you have to think *like* a system and attack multiple components simultaneously. We fall out of the habit because it's easy to revert to a new-product mindset or a productivity-initiative mindset. But the reality is that you have to go up a level for 99 percent of these ideas to work. If you liken this to the genome concept, you're motivated to look at the interconnectedness of things: you don't get the true power of anything until you attack using all elements in concert with one another."

Winter calls it a systems approach; we call it the business genome approach, a concept founded on the fact that, like humans, all businesses are made up of the same components. It's not the individual gene itself that makes the difference, but the combination of the genes in a particular pattern. That permutation is what will drive a company toward a new opportunity for growth or reinvention and an entrepreneur toward an entirely new playing field where he or she can land, hit the ground running, attract loyal customers, and grow.

Genomic Patterns Offer a Fresh Lens for Business

Learn from leaders of companies whose strategies reflect the Business Genome approach:

- Groupon, dedicated to bundling discounts online, took the idea of newspaper coupons and mixed it with the power of community to create a business that made $500 million in revenue in its first two years.

- OnStar matched the phone call home with GPS and the 911 emergency phone system to invent a way to exit the dashboard and enter Best Buy, morphing from a GM factory-installed service to an electronic device that can work on any vehicle.

- Cooper's Hawk infused great food and a relaxing atmosphere with Napa Valley vibe and added Facebook's viral sense of connection, community, and the power of recommendation to create a restaurant/winery that boasts the fastest-growing community of club members of any restaurant "cult" in the country.

- G.E. applied its desire of incorporating the new world of energy use and sustainability to its manufacturing powerhouse and invented its ecomagination generation of manufacturing products and facilities.

- Toms Shoes jumped on a growing philosophy to make a difference and defined a new space for shoes, in which the sense of community and social contribution could be married with every purchase.

The Business Genome approach is designed to allow everyone in business to follow their hunches, look at the world in a new way, and craft ideas for growth in just the ways that Groupon, OnStar, G.E., Toms Shoes, Cooper's Hawk, and Allstate Financial did. These ideas that became profitable organizations were conceived not by comparing an existing industry to the restaurants, insurance companies, shoe companies, or manufacturers of the company's competition, but by following a genomic process and finding patterns.

The Genomic Approach to Business

Businesses consist of components of great ideas that can be melded together to create and launch new business opportunities. When the higher-ups at Jiffy Lube started to suspect the value of selling oil had diminished in favor of a more holistic approach to the oil change experience, they used a genomic process to borrow ideas from other industries. Who had a great waiting room experience that could be built into a new model for Jiffy Lube? How could that experience be grafted onto the existing interior of Jiffy Lube's stores to create a killer app?

From Viscosity to Customer Experience: Pennzoil to Jiffy Lube

Jiffy Lube's road to customerization over the past decade highlights the shift in what matters now. Clyde Beahm, former president of Jiffy Lube during its early days of expansion, explains the complexities of driving customer loyalty by transforming the entire experience of an oil change. "Before, our options were limited and undesirable. We could either leave our car at a dealership waiting all day, or take it to the local 'grease monkey,' and sit in an atmosphere designed for dirty rags and tools, but not for customer comfort." Under Beahm's leadership, Jiffy Lube focused on the ambiance of the waiting environment, the speed of service, and the organization of the recordkeeping, driving customer loyalty for more than a decade based on those customer-facing priorities (Beahm, 2011). To keep up with the speed of change, Jiffy Lube realized it was competing with customer experiences well beyond the waiting area and billing procedures: its customer process for an oil change incorporated the proactive reminder systems of a computerized hair salon or dentist, the loyalty benefits of an airlines frequent flier program, plus the engagement of social media outlets where talking about top-of-mind topics like travel tips and fuel efficiency, gleaned from Internet feedback, involved customers in-between visits. Jiffy Lube went beyond differentiating itself as a comfortable place to wait for auto maintenance to a watering hole for chats with and between customers about owning, driving, and maintaining a car. It wasn't just oil changes anymore.

Beahm explains how the genomic approach became a catalyst for change. "Pennzoil could not have grown organically from discussions about viscosity to a vision for Jiffy Lube. No amount of benchmarking against industry peers would have gotten them there. It had to take a certain leap of faith that once it shifted to focus more on its distribution channel, it wouldn't be long before it was up against a new customer requirement: a pleasant oil change experience. No conversations about viscosity or formulations would have led naturally to that arena."

Pennzoil had to borrow customer experience insights from the likes of Starbucks to really create its next incarnation—Jiffy Lube.

The genome mindset sparks new insights for translating the DNA of one business to the growth challenges of another. It allows us to see through a new lens. Now we can read the facts and figures in a different way, and shift our focus toward response and away from prediction. What we'll need are ideas—novel ideas—influenced by strategies that have worked successfully in other arenas. The *Find Your Next* process can help us get there by offering a systematic way of sorting through those ideas to land on opportunities for building up any company, whether a large, global enterprise or a local start-up.

The heritage for the genomic approach truly began in 1991, when the Human Genome Project defined a new playing field for genetic scientists interested in the next step of DNA analysis, thereby enabling people to trace their individual biological inheritance. It was a huge project that started with gene mapping and led to new medical innovations, because, for the first time, scientists could actually look at patterns of biological information and visualize connections.

But genomics had much further reach. Along came Pandora, an online music service based on the Music Genome Project and sparked by the imagination of Tim Westergren, who, in 1999, decided that the time had come to organize music into new genres that freed us from the notion of old music industry categories like "rock," "classical," "heavy metal," "rap," "country," and "rhythm and blues."

The Pandora model started with a lot of trained musical listeners writing more than 600 descriptors of music. From there, Pandora.com launched as an Internet music service that combined the elements of music's DNA with a thumbs-up thumbs-down user voting that trains the station to refine its next selection for you. A personal profile develops and two things happen: one, you find yourself exposed to new songs and new musicians because the Pandora engine helps you match the taste of what you know you like (the Beatles) with music that has a similar "genomic" pattern (rock beat, good lyrics, lead guitar style), and two, you discover that your taste can't be categorized as simply "rock." You, and your preferences, are a little more complicated. The net result of listening to music with Pandora's music genome engine is that your musical taste expands to new genres, new styles, and new possibilities.

The Find Your Next approach builds on that same genomic concept. It is based on the idea that the possibilities for what a business leader can do next must come from somewhere other than what they did last. The Business Genome model isn't designed to help business leaders predict what comes next but to arm them when it happens. As was the case with Pandora's analysis of the world of music, all genomes are intrinsically human. What can a cereal company learn about a clothing designer's experiences with teen purchasing trends? How can a manufacturer benefit from a defense contractor's insights into applications of hydraulics? What can a medical equipment distributor learn from FedEx about global logistics innovations? Businesses are all about their customers, and that means they're all about people.

Vook: Plug and Play and Read

Brad Inman doesn't think of Vook as just another book reader, or what a customer views on Vooks as simple books with moving pictures. He considers the difference he brings to the world of traditional publishing to be genomic, in which his company's combination of talent plugs and plays into the new world of ubiquitous display and connectivity. Vooks, unlike traditional books, can tell their stories with movie-like engagement, a dose of mobile device speed, a bit of radio magic, the feel of a magazine serial, elements from the interactive world of Wii, and the sense of viral community popularized by Twitter.

Which combination of product innovation, talent deployment, process design, customer impact, secret sauce, and trendability can bring your company to a new level of opportunity? In that Business Genome vein, business leaders can diagnose where their businesses are in a way that goes beyond the traditional strength, weakness, opportunity, and threat (SWOT) tools. They can create a map of their company's current genomic pattern. They can look across industries for examples

A Personal View from Tom Stat: Learning to See Things through the Lens of Both Hemispheres

In my final year at Boston University, I expanded my electives beyond my degree path in psychology and discovered architecture. In architecture, I found what I believed to be the ultimate integration of art, science, engineering, and human behavior—a marriage between my two brain hemispheres.

This journey of integration and synthesis has remained a central theme in my life and in my work. After many fulfilling years immersed in the world of architecture, I joined IDEO, one of the world's most respected innovation consultancies. As a part of its laborious interview process, a good friend and mentor wrote a note to an IDEO partner. It said simply, "Tom was genetically engineered to work at IDEO." I had found a home where my genetic predisposition toward seeing analogies, leveraging unusual connections, extracting value from parallels, and creating something from what appeared to be next to nothing, were celebrated. Thus began some of the most rewarding, challenging, and fulfilling years of my professional life.

In almost 12 years at IDEO, beyond all the expertise, deep domain knowledge, design genius, and engineering prowess resident there, I believe that two true competencies of innovation were the primary forces behind our work and our success. Simply put, they were the ability of diverse disciplines to work collaboratively and the ability to synthesize across dissimilar domains. Crucial to both is the ability to integrate, recognize patterns, fuse, and combine disparate elements into entirely new entities. It is about leveraging a playful state of curiosity and harnessing the outcomes of discovery. It is the transmutation of lesser, more common elements into something of far greater value. It is alchemy in its purest form. It is nothing short of genetically engineering what's next, not merely near or new.

of companies that have tackled the same concern successfully. IDEO, a firm known for its innovation, especially in the area of product design, is one example. A few years ago, it was assigned the task of redesigning a hospital emergency room. Rather than simply benchmarking its client's ER against other hospitals, its management team studied another process that mirrored several components of a hospital's genome, one characterized by the need for precision movements when time was critical and the pressure of a potential dying patient loomed large. The

How to Open Your Organization to New Thinking

I once worked with a large manufacturing company that was trying to develop new chemical formulations, but its team was stuck in old ways. The company had a hiring policy it called the "nine-dot system," based on its recruiting policy for hiring the exact right fit from a competitor that had a nearly identical chemical development process. When it needed a new engineer to do job A at their company, it had the headhunters scan the competition for someone who did exactly that job. It didn't think to look for someone who could do job A and maybe job B.

Through what began as an experiment, the company's management changed that approach for one hire and brought in an industrial designer to fit into their team, thereby transforming the company's development process into a much more consumer-friendly approach.

To quote the client who was responsible for that change and the resulting double-digit growth that came from the new solvents, "If you can't change the people, sometimes you have to change the people."

A similar lesson came to an energy company that was a client of mine that was experiencing the early impact of deregulation in the 1990s. It tried to devise breakthrough thinking about how government regulations might change their industry and inserted a fresh perspective into its planning—by way of a new employee who came from telecommunications and had expertise in that industry's shifts. This new recruit brought some lessons about the FCC's telecommunications regulations into discussions about FERC's energy regulations and accelerated the company's adaptation to the change in their competitive environment.

answer for them was NASCAR pit crews, and IDEO grafted best practices onto the hospital's process redesign.

Taking a systems approach to business innovation requires some new skills. But that's the secret to discovery, to be able to shuffle elements to make a break through. We all have the capacity to think this way once we tune into business patterns and think about our futures in a new way.

Every company is capable of injecting the new perspective of another industry into its own DNA. The trick is to begin with a fresh mindset, founded on the belief that the future should get at least equal billing with the past as a catalyst for our organization's growth path. The beauty of applying genomics to business is that the business opportunities that emerge are part "art" or instinct (what-if thinking and a sense of what could be), and part science (analyzing other processes in unrelated industries and looking for patterns of opportunity). Only then, with a little of both, can companies get to a next with success that will stick.

Chapter
2

A New Outlook for Business

Recognize Patterns and Learn from Those Patterns by Adapting Faster to the Change You See Ahead

Pattern recognition is the first required fundamental or skill in the genome transformation process. Business leaders must learn how to spot them within the business universe, look beyond their company's own industries for insights into their company's future, and interpret what they see.

G.E.'s recent innovations in transforming the corporate culture offer powerful illustrations of exactly that: how a company can go from recognizing genomic patterns that could drive its strategic priorities to making an actual corporate shift. The process began a few years ago when Jeffrey Immelt, G.E.'s chairman and chief executive, started exploring ways to retool for the future, looking primarily at companies in other industries for some inspiration. These companies had much to teach him in the two components—innovation and discipline—that could best feed G.E.'s hunger for rapid improvement. Immelt found those essentials in Google's internal entrepreneurism and West Point's leadership culture of adaptability, and then he grafted them on to G.E., which led to dramatic shifts in its corporate culture toward a more team-based, innovative structure.

Whether large or small, the take-home message is the same—looking at today's patterns of success in multiple industries fuels tomorrow's ideas for what could be.

The Art and Science of Figuring Things Out

There is an art, and a science, to figuring out our next.

How do companies interpret the same facts that everyone sees but in different ways?

Before we can begin the process of sorting through our options, we must first know where we are and where we're going. It isn't just a matter of looking, but interpreting. After all, we see what everyone else sees. And once we understand our observations, we can't stop there; we need to also teach ourselves to think differently about the data we analyze, the questions we ask ourselves, and the conclusions we draw as we imagine new future directions.

There is no science to prediction. You can sit down in a laboratory with rats for a month and track what you see. Even armed with data on how biological change has occurred in the past for each rat, you wouldn't be able to figure out what changes will happen next. And why not? Because you have not been trained to see the world of rats through the subtle clues and environmental shifts that would allow you to see ahead to the next phase. In other words, there's no way to know for sure.

Business is not much different. Sit down with a mountain of forecasting data and you would face the same problem. Because you're not trained in evaluating the potential opportunities lurking behind new trends, it would be tough for you, as a person running a company or even a division of a company, to imagine which move would give you an edge in a changing competitive environment. You can't know all that lies in the periphery that puts things in motion, and places certain companies ahead while knocking others behind. Most business leaders haven't focused on responsiveness or rapid adaptation; they've been wired to analyze what was and what is. But companies that thrive today have developed a new skill: a talent for interpreting facts more insightfully and tapping into the non-obvious. FedEx did exactly that when it transformed "speed" into a point of difference in package delivery. Lululemon, too, did it when it saw a nation of decentralized yoga enthusiasts and gave them a brand to embrace.

Teaching Our Businesses about the Wetsuit Moment: When It's Time to Respond with Insight

The skill that scientists use to interpret the chaos is the same skill that drives success for FedEx, Lululemon, and most other companies that stake new claims in the competitive arena. The leaders of those organizations have mastered the ability to not only navigate the same waters everyone else can, but to see early signs of change and adapt quickly. With greater peripheral vision, they have become proficient at seeing more than their fellow snorkelers. They see what lies ahead. In essence, they come equipped with a "corporate wetsuit" and put it on when the time calls.

If a snorkeler is swimming in warm waters, but, on the horizon, suddenly spots an igloo floating on a slab of ice, what should he do? Should he take it as an illusion, or put it out of his mind? Or interpret it as he should, as a sign of colder waters ahead?

He can put on a wetsuit and move through it, or he can do nothing and freeze to death.

Could it be that your company is luxuriating in today's warmer waters, and ignoring opportunities—or risks—presented by the igloos of trends just ahead? We all need a basic reality check.

Your business needs a wetsuit if:

1. You're starting to see some disconnects between customer satisfaction scores and sales. (They say that they still love you, but they've stopped buying.)
2. You're reading news about businesses that do something sort of like what you do but are branching out (i.e., Blockbuster and Netflix).

3. Your pricing has become a commodity—people aren't willing to pay a premium for the same things they used to.
4. Your competition is innovating and you're not.
5. You introduced a price increase that led to higher top line revenues at first, but now customers are disappearing.
6. There's a new world order that could have a huge impact on your company (recession, globalization) and you haven't been hit with a wake-up call yet.
7. Everyone in the company culture is more committed to "that's how we've always done it" than "that's how we're keeping up."

Moving from Seeing to Doing: Being Nimble

Recognizing change on the horizon isn't the same as reacting to it—or even thinking ahead of it. To drive our organizations toward the opportunities that lie ahead, we need to take our existing businesses down a new path or tap into the potential we envision. But that requires mastery of entirely new skills. The killer skill for today is nimbleness—not necessarily having the perfect model on Day One or the perfect read of every fact at the outset, but the ability to change direction quickly when the path isn't leading to success, or when new competition, economic factors, or other forceful conditions get in the way.

When P.F. Chang's thought it wanted to expand a few years ago, the company first tiptoed into an extension by introducing a sister concept, Taneko Japanese Tavern. It didn't work. What saved the company from suffering too dramatically from that decision was its corporate muscle for adaptation: it was nimble enough to shift gears and try two others that did work (frozen meals and global partnership) and it was better off. If it had clung too emphatically to SWOT analysis or forecasting models, it might have missed the boat while the other boat sunk. Instead, it read the future quickly, recast its net into new waters, and accelerated toward better concepts. (Read more about how P.F. Chang's turned things around on page 145.)

Stories like P.F. Chang's underscore what every business leader faces today—the new forces working against us as we try to grow. We are all facing new realities: the mountain of facts is huge, the speed of change is impossible to keep up with, the information that used to keep us ahead of our competition is now instantaneously available, our customers are talking about us to each other more than ever before, business dynamics have turned global, and the expectations for competitive advantage are rising at record speed.

Yet, just like P.F. Chang's, we are expected to plan for the future, innovate, and predict what's next, with yesterday's tools. Like generals asked to fight a war in a new jungle using Napoleonic battle tactics, we're equipped, not with bayonets and lines of troops, but with bows and arrows that used to work but don't any longer. We need to arm ourselves for our new world. We need to form a new approach to seeing the future. We need to be less concerned with a desire for certainty and move our focus away from static dashboards. And, when all is said and done, we need to spot the needle in the haystack. We need to find a unique idea and run with it.

The One-Two-Three of Competitive Edge: Speed, Crowd, Globe

It can be overwhelming enough to bring an organization from its comfortable state of status quo to a vulnerable state of change. But to succeed, that's what we must do. And that means always keeping a steady eye on the horizon. But what about what we can't see until it's too late? We've all been surprised when a trend in our rear view suddenly emerges and makes itself known. But, there are three factors that everyone should use as their preflight checklist before they explore the potential of a new direction:

1. **Speed**. Amplify your organization's ability to keep up with instantaneous cycle times and the accelerated speed of commerce. FedEx based its whole enterprise on speed; JP Morgan Chase's mobile check scanners added lagniappe to itstraditional financial services; Vook reduced the cycle time to produce a book by months with its multitasking multimedia platform. Every company should consider whether it can use speed as a marker of differentiation in some aspect of its vision for growth.

2. **Crowd**. Acknowledge the shift in dynamics from a past when companies could telegraph a one-way message to the masses to a present when commerce is being crowd-sourced. Customer conversations about our companies have gone public, instantly available and readily accessible in blog posts, forums, online groups, mobile messages, and shared links. That doesn't mean every company has to choose to center its entire business strategy on the crowd. But, every company has to recognize the crowd as a new force and learn how to integrate the power of many into its next steps.

3. **Globe**. Some companies that have long been known as domestic-only brands have recently stepped onto the world stage to expand their reach. Other brands that center their entire enterprise on global connectivity have made that fact even clearer.

New Times Call for New Tools: Take Out Your Periscope

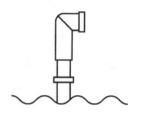

It's time to retool our companies. Strategies that overemphasize our past results—depending too heavily on the rearview mirror—can result in our having torearrange deck chairs on the *Titanic* while; missing the big icebergs ahead.

The microscope point of view is no better. If we rely too much on examining our current financial results and metrics as a way to discover our next opportunity, we run the risk of assuming that our future will be like the promise from the P.F. Flyers sneaker ad of the 1950s, a "run faster, jump higher" version of today. It could well be that a new trend is staring us in the face that will require a more disruptive or dramatic strategy. As Adam Kanner said, "ScoreBig had to be built by a third party in order to service the entire industry. Organizations like the NBA or the NFL could never have done it—they're too big, too focused, and only represent a fraction of the sports and entertainment industry as a whole. There needed to be an agnostic, arms-length, industry-wide solution to this huge industry and consumer issue of ticketing, and that was the opportunity for ScoreBig." But, that's exactly what some companies need—that bigger strategy—to change course.

Sometimes the gravitas of a forecast model—the telescope—needs to be balanced with a new view to inspire a company to move nimbly in a new direction. A large energy company I once worked with invested heavily in scenario planning and forecast modeling, only to be too slow to react when the Gulf of Mexico oil spill happened. Large financial services companies that depend solely on forecast models still didn't see the 2008 Wall Street meltdown coming, and they were slow to react when they did.

Forecasting is rarely enough. These traditional tools don't adequately address the three significant trends that shape today's competitive arena. What is needed is a new tool, the periscope, that allows companies to see the future today, in the successes of entrepreneurial companies entering our customer's preference lists (like Flickr's entry into the world of photos), the innovations of large companies shifting their logistical models and product offerings (like P&G's Swash "pop up stores" on university campuses and their dry cleaning franchises), in global competition (like Johnson & Johnson's insights on the potential of European waterways as a way to speed up logistics, borrowed from Nike's warehousing and distribution models), and in the tea leaf patterns of game-changing trends that are in plain view (like personalized medicine). With a periscope, we can take the risk out of guessing what might come next and stop crossing our fingers hoping that our forecasts might come true. Take out yours and map your company's genome to opportunities on the horizon today, graft innovative successes onto your business, and adapt quickly to the change you see.

Address the new realities and achieve cha-ching in the new era:

1. Read the future today. Start by interpreting the genomic patterns, look to industries other than yours, and build new awareness about what your organization's information is saying to you today about tomorrow.

2. Keep up with insanely fast cycle times. Exercise your organization's muscles to be nimble and to adapt to the change you see.

3. Enter the crowd dynamic wisely. Re-tune your company's people skills. Listen to the themes, participate in the conversations, and engage with the community of people whose opinions can drive your company toward a new area.

4. See the world. Insert the global perspective into your localized thinking. Whether yours is a large, global company or a small, local business, challenge yourself to bring the world view onto your radar screen.

5. Transition from SWOT thinking (strength, weakness, opportunity, threat) to Find Your Next or Business Genome thinking (sort, match, hybridize, adapt).

The next chapter will introduce the tools to get started.

Chapter 3

Find Your Next Process

Find Your Next Steps to Next

1. **Sort** through the options.
2. **Match** your genome using the Business Genome elements
   a. Product and service innovation
   b. Customer impact
   c. Talent and leadership
   d. Secret sauce
   e. Trendability
3. **Hybridize** your company.
4. **Adapt and thrive** by breaking out of old habits.

Our business environment has changed, which means we have to change just to keep up. Our world is now crammed with data and we must mine it precisely and with great care to get to the facts we need. We're well-armed with metrics, reports, and very good intentions. But we might run the risk of tipping the scales in favor of an incremental, predictable approach to our future, even when the greatest opportunities might not emerge directly from our past. Do our current spreadsheets and four-color graphs and animated PowerPoint presentations lead us down the path toward the most powerful future of our own design? Perhaps not.

What about the questions we should ask ourselves as we dig into all that information? What about the options that our models don't clearly show? What about our instincts? What about the ideas we need to see emerging from opportunities, the ones that don't come from piecing together last year's results into a better-performing version of the same thing over and over again?

On the one hand, there is expansion on our "now"—organic forward motion that builds on our current core strengths, logistical evolution based on improvements in efficiency, and fine-tuning of products and services to move us up the ladder in the minds of our customers—and on the other hand our more novel "next"—product development based on new technologies, and leapfrog innovation that strikes a new chord with the customer groups we haven't reached yet. It is time to stop looking at the future with a one-size-fits-all toolkit and realize when our organizations have moved beyond their "last" and push them on a better track toward their "next."

We must get to the heart of what will take us where we truly need to go. But how? How would you get started? What filters would you use?

The traditional way was well-suited for the moment when our companies needed to keep doing what they'd been doing, only at a faster speed, with higher efficiency, or with slightly different offerings. But, for most of us, today's business environment has changed in fundamental ways that require us to master more than process improvement, product enhancement, and competitive advantage. Today, we need to shore up to our *responsiveness* and embrace a less-traditional model, then equip ourselves to be top-notch at seeing opportunities on the radar, and finally flexing to the speed of change.

The responsive approach to strategic planning doesn't start with a completely blank slate, a unilaterally innovative mindset, or a complex set of what-if scenarios. In the world of commerce, creativity—something we need—has to be judged by results, financial strength, shareholder value, customer loyalty, and market traction. *Cool* has to fuel sustainable growth. The pull of the new or next shouldn't replace or supersede the power of what is working now for your company. Most organizations don't need to abandon ship, but, to a certain degree, they can get better with moving along. For those companies, the steps to next can be implanted into their overall portfolio of options, strategies that build on the now (improving efficiencies, expanding incrementally) and embrace the next-door (market expansion to adjacent markets, aggressively capturing market share) and the *Find Your Next* process.

The refreshing difference is that when companies look for their "nexts, " they now have a new point of departure—the future.

Find Your Next Process

The *Find Your Next* process is designed to balance the strength of traditional models with the flexibility of intuition. With greater speed, nimbleness, and creativity, we can capture the dynamics of today's changing business world and make them work for us. The steps are designed to allow us to embrace the ambiguity of what lies ahead, follow our hunches to new opportunities, and coax ourselves into stepping outside our traditional industry comparisons to grab successful ideas from outside our comfort zone.

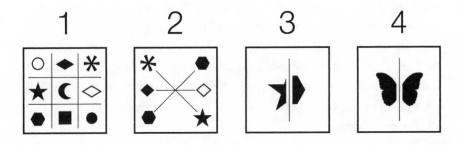

This four-step process provides a road map. It's not so much that the questions you'll ask yourself are new, but that the starting point has changed: it's away from strategic processes that are balanced in favor of the past and toward a mindset that leans in favor of the future. And the emphasis is not on perfect end points for forecasts—the muscle to flex is that of adaptation—but responsiveness to the change we see while creating the products, processes, customer approaches, and culture that will take us where we believe we can or need to go.

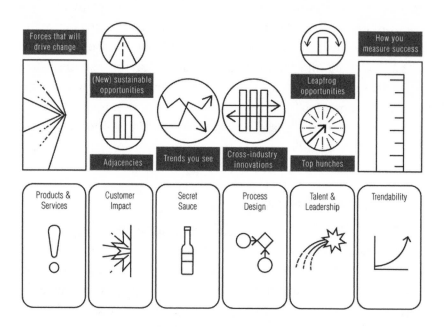

You'll also see that the data points on the competitive screen don't simply rely on our own industries for comparison; benchmarking is only part of the competitive landscape. Using peripheral vision to see into markets other than our own is just as important as looking straight ahead at the competitors in our direct line of sight.

The questions we ask ourselves might seem familiar: Where do we find our next opportunities for competitive advantage? How do we anticipate the forces on the horizon? How do we stay front-and-center in the eyes of our customers today and tomorrow? How do we keep up with the pace of change?

The process has changed. We're looking to excavate hunches, take the pulse on the future based on the pulse of the present, and integrate some of the foreseeable new forces (competition, adjacent fields, and unrelated fields) into our vision for tomorrow.

Answer the following for yourself:

- Are you at risk of becoming obsolete? Are you facing a shift in your market?
- Are you off-trend?
- Do you have a hunch that there's a new direction you should pursue?

Follow the four steps. Ask yourself the questions. And rethink your next.

STEP 1: Sort

The process begins with an honest description of your current genome (the six elements that make up your business) and the ways they combine to enable you to differentiate yourself from your competition and engage your customers. You can take each element and ask yourself: What is working well with this aspect of our company? Where are we

The Liar's Box: Wake-Up Calls for Companies

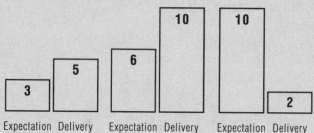

The year was 2008 and a Swiss-based engineering firm had just won an award for the sharpest drill bits in the industry. Its machining was rated a 5 out of 5 and its customer service scores were through the roof. Its brilliant designers received bonuses and accolades. The company had a cost-effective process for delivering the product on Thursdays of every week, a decision it had come to through process optimization based on bulk shipment rates. Then, the problem arose: sales started to slump. By the time the company called for help, its market share was double-digits down and it was at risk of being over taken by a competitor.

The company was suffering from the syndrome that Tom DeLano, principal with Market Metrics, calls the Liar's Box, in which surveys and analysis tell us one story with certainty (the drill bits were perfect and the customers loved them), yet the numbers tell another (the sales were dive-bombing and market share was sinking fast).

In the case of the Swiss company, it neglected to ask a very important question that went beyond the engineer's mindset: "How important is it to receive the spare part on a given day?" It turned out that by not asking that critical question, all of the engineering in the world wouldn't earn it favor with the customers running oil rigs that cost hundreds of thousands of dollars per day to operate. The supplier's focus on the drill bit design and cost-effective bulk shipping didn't match the expectation of just-in-time delivery (something that the competition had mastered, despite a slightly less perfect drill bit design). The customers couldn't afford to lose half a million dollars in down time for the rig if they needed a part on Sunday that wasn't delivered until Thursday.

So, the drill bit designer had to uncover its Liar's Box—that illustrated it was kidding themselves into complacency—to focus on what mattered most to its customers in order to recapture market leadership.

falling short or falling behind? What's the combined effect of our company's impact on our customers now (with all of the six elements combined)? Do you have a hunch about where your company should go next? Is there a metric keeping you up at night? Do you see early signs of change ahead? Which trends are affecting your company right now: faster cycle times, a new customer dynamic, and global forces?

Sort Through Metrics

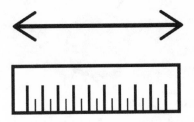

The second piece of the first step is to assess the dynamics of your market or the shifts in your competitive environment. Ask yourself if you want to keep providing the same products and services to the same customers and clients. If so, what forces do you have to take into account to refresh your outlook and your approach? What trends could have impact? Are there early warning signs that the landscape is shifting? Have your margins started to shrink? Are your products becoming a commodity? Is someone else doing a better job of serving your current customers than you are? Are you considering offering the same products and services to different people? Are you considering offering different products and services to the same people? Do you have an idea for an entrepreneurial concept that will break new ground?

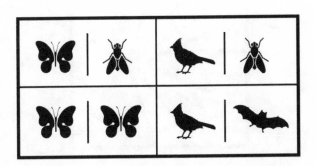

Are there signs of opportunity in an adjacent field?

Is the overall market growing? How can you capitalize on the wave of growth?

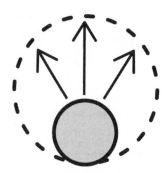

How would you capture a bigger piece of a market that isn't growing?

Can you envision a new opportunity that is sliding in next to your current sweet spot?

Can you lead your horses to different sources of water? Is there a "white space" opportunity for your company that taps into new trends or new customer appetites?

Sort Through the Dynamic Context

Now take a look at the dynamics now at play.

- **Forces.** Which forces do you think will reshape the future? How have you prepared for them?

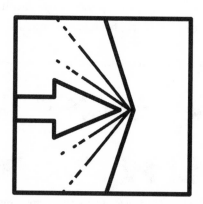

- **Trends.** Ask yourself the hard questions, and answer them: How have you responded to the new realities of speed of cycle time, hyper connectivity, viral communication, and global dynamics? What other trends are about to change your world? What portion of the total dollars spent is being spent on *your* offerings? How are patterns changing? Where could you insert your product or service potential in the new landscape? What are the value drivers for the next customers? What matters most to customers now? How will that change in the future? Is your business due for a sea change?

- **Opportunities.** Do you believe your company could "leapfrog" beyond its current approaches and succeed in a new area, with new customers, or with new technologies? Which new technologies, distribution channels, products, or offerings could accelerate your progress?

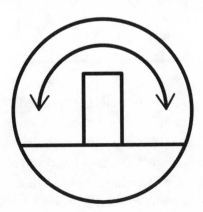

Sort Through Your Hunches

What general hunches come to mind about your core genomic elements? Where do you think you stand? What do your instincts tell you about how you might recombine your core elements to get ahead of emerging trends?

STEP 2: Match Your Genome

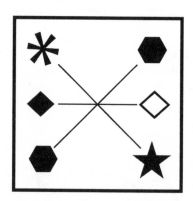

Now that you understand yourself, get inspired. Take your business genome and match it to the genomes of others. Dig into other industries and organizations that have cracked the code. Expand your mind. Don't think industry-specific, think focus-specific. If you're looking for a better way to attract top talent, look at the companies that have been voted best places to work: SAS, Boston Consulting Group, Wegmans Food Markets.

If you're trying to smooth out a complex process, study the masters who have already figured it out. If you'd like to expand your product business to include service, look at who else has created a customer-friendly experience (Zappos). If you need to construct better mousetraps or build out your viral platform, look at the most innovative companies for ideas. (For starters, check out Fast Company's 2011 list: Apple, Twitter, Facebook, Groupon, even Nissan.)

The power of the genomic approach exceeds the traditional definitions of "best practices," though. As we've said already, it's the combined effect of these elements—their genomic relationships—that bring power to them as strategies. Look beyond the individual components and consider the patterns and integrated effects that make things work. Don't just ask yourself, "How can we graft that foreign element into our corporate culture?" Instead ask yourself, "What would our organization have to do to make sure that graft fuses successfully and helps our company grow?"

Find Your Products and Services

Dig a little deeper into each component. Have your products run their course? What would their new course look like? What type of innovation could lead to an uptick in sales? Where is the value in what you have to offer today? What companies in adjacent competitive areas are starting to erode your market share? What companies in other industries, other geographic areas, and other markets exemplify the type of product innovation that would get you to your next?

Find Your Customer Impact

Which part of what you offer resonates most with the customers of today? How are customer preferences shifting? What aspects of the customer perspective should you pay more attention to? Is there a latent or unmet customer need you could tap into? What would the ideal crowd-based conversation look like? Where is customer taste heading next? What companies have the type of customer impact that would get you to your next?

Find Your Talent, Leadership, and Culture

What talents do you need to integrate into your organization so you can jump forward? What type of culture do you believe will drive or support competitive advantage? What is the profile of leaders for the future of your company? What companies exemplify the approach to talent, leadership, and culture development that would get you to your next?

How well-suited is your organization's culture to adaptation?

Find Your Processes

How could your logistics shift to get to you to the future faster? What should the new givens be in the area of processes as you move forward? Which operational components would you review, streamline, shift, update, or rethink to improve your game? Which companies exemplify the approach to processes that would get you to your next?

Find Your Secret Sauce

Do you have a secret sauce? What is it? Is it—and your brand—still relevant? What brands have begun capturing the attention of your customers? What are the killer apps? What would push your company beyond incremental innovation? Where are you at risk of becoming an also-ran, commodity, or even worse, obsolete? What are the elements that would transform your brand from lukewarm to hot? Which companies have the types of brands and competitive distinction that would get you to your next?

Find Your Trendability

What would your company look like if it were integrating trends more effectively into product development and operations? What companies demonstrate the responsiveness to trends and nimbleness that would get you to your next? Which trends from other industries might have an impact on your company's future? How is customer taste evolving? Which priorities are becoming most important? How is your company adapting to the speed of the market? How is your company responding to the voice of the crowd, social media, crowd sourcing, and transparency of web-based communication? How is your company embracing the dynamics of the global market? What trends are you missing that others haven't?

STEP 3: Hybridize

The third step involves cross-industry grafting: finding a genomic pattern that works in a company from another industry and grafting it onto your own to accelerate growth. If you are an insurance company looking to build your e-commerce, don't look to other insurance companies like Geico or Progressive. Instead, look to the giants of online commerce like Amazon and Zappos and customize their process to fit your structure.

The Cross-Industry X + Y = Z Formula Is in:

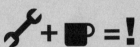

P.F. Chang's China Bistro + HOV lane = Pei Wei (fast casual dining option)

S+H Green Stamps (newspaper coupons) + Facebook = Groupon (viral, online promotion of product/retail discounts)

ER + drug store photo service = Minute Clinic (fast, neighborhood, streamlined medical care)

Procter & Gamble + Greenpeace = Method (environmentally friendly detergents and soaps)

Shoes.com + UNICEF = Toms Shoes (when you buy a pair of shoes, Toms Shoes donates a pair)

Sundance Film Festival + 1000 Points of Light = IndieGoGo (democratized crowd-funded support of independent projects)

Pfizer + Procter and Gamble = Pfizer's Global Centers for Therapeutic Innovation (decentralized research and development based on P&G's "Connect and Develop" philosophy)

Red Cross + Text Messaging = Haitian Disaster Relief (public support of natural disasters via text messages that collect donations on a huge scale instantaneously)

Caterpillar + NASCAR = Lamborghini (started out in farm equipment and transitioned to sports cars)

Grey's Anatomy + Disney = Sharp HealthCare (took Disney's "on stage" versus "backstage" approach to redesigning the entire experience for employees, healthcare workers, patients and visitors)

Armani Uniforms + Paris Hilton = Armani Exchange (started out as military garb, transformed into fashion)

Brooks Brothers + Avon + Dell = J. Hilburn (men's suits sold by Avon's word of mouth, created just-in-time, and customized like Dell computers)

Fighter jet + Helicopter = Harrier (a fighter jet that can hover and land on an aircraft carrier or in jungles without landing strips)

Vending machine + Google maps = Placecast ("geofencing": the ability for a customer to opt in to a text message service that customizes your likes to your location)

If your company were able to adopt characteristics from another organization in a different industry, what would they be? Think outside the box, or outside the industry.

STEP 4: Adapt and Thrive

The final step teaches your organization to adapt and thrive, and deprogram yourself from trying to predict what's to come. The future is here, now. Embrace the new strategic realities. Combine the elements so you have what it will take to grow your business, and look at what already exists. Recognize the shifting competitive environment, borrow successes from other companies and other industries, and jump ahead of your competition.

A Word on Context

The *Find Your Next* process is not just for when your company needs to completely revamp itself. The Find Your Next process is designed so you can significantly enhance your current offerings through the inspiration of other industries, serve new customers with what you currently offer (quadrant 2), or offer new products and services to your existing customers (quadrant 3). In some cases, when breakthrough transformation is the best way to next, you can use the process to uncover latent needs in the market, tap into emerging customer preferences, and devise an entirely new invention or service (quadrant 4).

Different offerings Same customers QUADRANT 3	Different offerings Different customers QUADRANT 4
Same offerings Same customers QUADRANT 1	Same offerings Different customers QUADRANT 2

But you can't begin looking for inspiration until you know where you are. Decode your DNA to find out just where you stand in each of the six genomic elements. Go through the *Find Your Next* process to next first.

Read on to see how each genomic element is broken down and explained. Every company faces fundamental changes as they attempt to evaluate their own genomic patterns. Read through the next section to better understand the genomic approach and how it manifests in real success stories. Just as the human genome in biology is broken down into G (guanine), A (adenine), C (cytosine), and T (thymine), the Business Genome elements can be combined to describe the DNA of a particular organization.

Although each element acts as an independent lever that can be moved to enhance a company's performance, the real power comes from using a few of the elements in concert with one another. Only then can you get from where you are today to where you want to be tomorrow.

Part 2

THE BUSINESS GENOME ELEMENTS

4

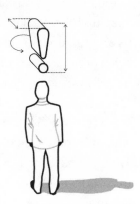

Product and Service Innovation

Manufacture Your Company's Next: The Secrets of Innovative Product Design + Market Analysis

Your Product and Service Innovation Self-Diagnostic

1. The world—and the customer—have changed. How is your stuff keeping up?
2. Where do great ideas come from?
3. The environment of the marketplace is complex and dynamic. What factors are at play in the game of selling?
4. Everything keeps changing—or does it? What genres of products and services are guaranteed to excel, and stay that way?
5. Some companies seem to have figured it out. What, exactly, have they figured out and what can you learn from them?

The world of business used to be easier. You started with what you were good at, you made sure people wanted it, and you put it on the market. Not so anymore. Selling has gotten trickier. Not only are we—the customer—more demanding, but our needs now regularly change in a capricious way we barely understand ourselves. With the donning of such a mercurial temperament we put an end to the days of assembly-line planning and traditional manufacturing plants that can no longer keep up with us. Before a new world of glimmering electronics seduced us and retrained us, a company like Ford Motor could follow a cut-and-dried process of number crunching by simply taking a count of the number of households who needed a Model T, its only real worries being time-sensitive customer surveys and large scale manufacturing. Today, there's no way to know if what we want now is what we'll want later, when a product finally makes its way into stores and our line of sight. That classic question—if you build it, will they come?—has suddenly become a lot more relevant.

Innovation, as a result, has revolutionized business, and the word itself has become far more meaningful. We've been wowed by its impact on our companies and our consumers. It's the phenomenon that has made Technology Entertainment Design (TED)—a menu of live events that feature innovative speakers on topics that encompass each genre—so successful. Every TED talk is posted on the Internet and combined, the online videos have attracted more than 250 million viewers to date. According to Chris Anderson, TED's curator, today's environment for innovation has accelerated partly because of the new tools and new gathering places for our ideas. He traces the timeline of the development of these virtual watering holes farther back than this century alone: from the trade routes in 3,000 B.C., to the sixteenth century's scientific revolution and widely distributed print publications, to the twentieth century and computing advanced code creation through open source, highly interactive collaboration, to today. Now, with video cameras on cell phones and easy-to-use software, we've raised the curtain on ideas by enabling people to upload videos to YouTube, create original programming, and get instant feedback, approval ratings, and a ready and willing fan base through the revolutionary vehicle of social media. With Facebook and Twitter, our ideas can now travel at record speeds (Anderson, 2011). Listen to Matt Ridley explain the power of the phenomenon in his talk, "When Ideas Have Sex." He presents a compelling case for the importance of interaction as an accelerator of creativity and the advancement of ideas (Ridley, 2010).

In business, the power of collaboration to advance ideas more rapidly can function as a double-edged sword. The days of resting on old laurels may be over for any company whose raison d'être was centered on a slightly cooler innovation, like the pre-Bloomberg makers of the old-school Wall Street tickertape machines. But, then again, if we refocus ourselves—toward weaving new realities into the fabric of our companies—and veer away from protecting our one "cool" idea, we can change with the times. Whirlpool did it when it transformed its 1948 roots from a classic, basic appliance company into a customer-focused innovator with eco-friendly options like quick refresh steam cycles. Innovation helped reposition the standard into a responsive, customer-oriented company that now earns the thumbs up from its fans as maker of "the best washer I've ever owned."

Bubble Ball: The Young Adult Genre of Entrepreneur

The world is becoming more business friendly because the Internet acts as an instant marketplace. Just about anyone can add a product to the mix, particularly in the virtual world. Bubble Ball is one such example, created by a 14-year-old on his iPhone. In one week, Bubble Ball was the most downloaded free app. And its inventor isn't done. He plans to release more games, all while he's still in the eighth grade. In his next one, he's considering for-profit options, because, in today's world, anyone can become an entrepreneur.

We've accepted the new truth that without a lot of better, faster, smarter, and sometimes cooler, ideas, we'll be wiped out by the competition. But what is cool? Cooler for a generation of customers entering adulthood in 2011 and beyond means something very different than it did even just a decade ago. In a world where five-year-olds download apps, and anyone can see an idea, pass it along, improve on it, and take something to market in record time, being creative isn't good enough. This world has new rules. Just look to gaming, as described by Steven Johnson in his book, *Everything Bad is Good for You: How Today's Popular*

Culture is Actually Making Us Smarter (2005). By zeroing in on the video game *Zelda*, Johnson underscored the transformations that have taken place just since 1987: a past of joysticks and remote controls and rudimentary graphics to a present of multi-player options and complex interactive programming that immerses the player into a much more creatively conceived gaming world. Fast forward to the post-Avatar world, and that same player—thanks to a global Internet-based multi-player platform, mobile technology, and a host of other innovations that are part of today's basic toolkit for gaming—can now live in a whole new virtual reality.

Transfer that technological mindset to the world of business and embrace the possibilities, because, as Johnson says, it does no good to resist: the world has changed and it's time to tap into the power of the new cool. The rules are simple to list, but it's tough to persuade an organization to follow them. Companies like Whirlpool had to make a large-scale, conscious commitment to start the ball rolling in 1999, and it took them almost a decade to earn the "best washer I've ever owned" moniker.

The Rules of Cool

1. Cool is happening in the mainstream (i.e., Whirlpool) but also on the fringes (i.e., YouTube).
2. Cool is an ongoing, fast-moving phenomenon, so companies can't wait for an annual strategy session to innovate.
3. Cool to us isn't always cool to our customers. But, we can monitor our customers' evolving tastes, purchasing patterns, opinions, and unmet needs. Social media and online analytics speed up the cycle time for feedback from weeks to minutes.
4. Cool is related to financial results. Early warning signs of *uncool* can be flat revenues or lower margins, red flags that your innovation is lagging.
5. When it comes to business, cool isn't just "brainstorming" or "ideation"; filling up walls with sticky notes isn't the end game. Cool is about figuring out *which* sticky notes will resonate with customers on a sustainable basis. Don't do cool for cool's sake.

To see how far product innovation has come, take a look at the history outlined in Allan Apuah's book, *Strategic Innovation: New Game Strategies for Competitive Advantage* (2009). Product strategies met business challenges, and as those challenges changed, so did the corresponding product solutions. The '60s focused on growth, the '70s on diversification and portfolio planning, the '80s on competition, and the '90s on competitive advantage. And today? For companies and executives that grew up in a time of classic strategic planning, trying to keep pace with a marketplace that won't stand still isn't easy. In fact, it's disorienting. An accurate snapshot of where we are today and what will hit us in the face tomorrow doesn't really exist. We can't know for sure what the future will bring. Or how to innovate to allow for what it does. And guessing is becoming more and more difficult. How are we supposed to plan? How can strategic models that depend on the ability to hold many competitive forces constant possibly prepare us for a competitive landscape that keeps unfolding? And what does that mean for what we bring to our revolving marketplace?

We're living in a different world. Which means that *cool* has changed the dynamics of what keeps our products relevant. And we have to be on the lookout for companies lurking on the margins of our businesses, poised and ready to grab customers, patients, clients, and guests right out from under us.

Walgreens: Taking Care of Customers and Patients

Walgreens, a company that transitioned from a pure pharmacy to an onsite health provider, now performs as many diabetes tests in one year as a traditional doctor's office with its Take Care Clinics. In a similar vein, brick-and-mortar college campuses can be outpaced, and in some cases replaced by, online universities. Case in point: University of Phoenix, a computer-based institution that focuses on self-paced coursework, took students out of live classrooms and left them at home, enrolling upwards of 345,000 students in 2008, making the segmented university roughly the size of the combined campuses of the State University of New York.

We need to adapt. And when it comes to product design, that means a piece is missing from our toolkit, without which we cannot bridge the gap between how we were trained and how we weren't. Just thinking about it will get you nowhere. And denying its existence won't fare you any better. Face it: this era of innovation is here to stay, and the bar for inventiveness will only continue to get higher.

The late '90s and beyond marked a time of heightened awareness of the challenges facing business leaders. Thinkers like Clayton Christensen, author of the 2003 book *The Innovator's Dilemma*, and Tom Kelley, author of the 2001 book *The Art of Innovation*, entered the scene and introduced the concepts of innovation as a mindset and method of sustainable product inception. The books became references for a new day and age. *The Innovator's Dilemma* included case studies of companies that created breakthrough, transformational products, emphasizing what Kelley calls "disruptive technology," or the art of uncovering unmet customer needs, and *The Art of Innovation* provided insights into the workings of the author's firm, IDEO, and its unique blend of creativity and observation.

Kelley's message was designed to shift thinking between yesterday's mindset—that only creative people can create product-driven business solutions—to today's—that all business leaders can create product-driven business solutions and have to in order to keep pace with the complexity and speed of the new competitive arena. Following Kelley's rules for innovation, companies could observe their customers with greater sensitivity, and recognize the inherent solutions to product development problems that were solvable by looking at another, unrelated industry. Kelley cites numerous examples of innovative product solutions that marry an element from industry A to a problem in industry B.

IDEO's Problem-solving Approach to Building a Thin, Lightweight Laptop

"In designing the slim Dynabook laptop, we discovered there wasn't enough room for traditional fasteners. A brainstormer generated the idea of 'knitting' the two halves together—like a piano hinge—which sent us out looking for a strong, thin rod threaded on one end. We eventually reached a solution by getting our favorite Palo Alto bike shop to supply custom spokes to Dynabook that, amazingly enough, did the trick" (Kelley, 2001, pp. 62–63).

The ocean swept in in 2005, a concept birthed by W. Chan Kim and Renée Mauborgne. In their book *Blue Ocean Strategy* (Kim and Mauborgne, 2005), they explain how to swim beyond the "red ocean," a place where competitors circle like sharks, to the "blue ocean," a less crowded, less threatening space. Kim and Mauborgne profiled organizations like Cirque du Soleil, a marvel that refuted predictions that live entertainment was on the decline and was being replaced by video games and other new electronic forms of leisure.

> Cirque du Soleil demonstrated what Kim and Mauborgne coined a "blue ocean strategy," a place where people weren't playing—the world of innovative live circus arts—and thrived. And it has continued to do so. Cirque du Soleil has expanded its performances to 271 cities in 32 countries, with plans to expand even further (Collins, 2009).

Kim and Mauborgne—through their blue ocean examples—advanced the strategic agenda by looking very innovatively at a company's potential playing field. From a practical perspective, business leaders should not only be innovative, but strategic, and integrate the best of these methods with the analytical rigor of traditional strategic planning to formulate an integrated process that keeps pace.

So what have we learned from those business thinkers at the forefront? We can count on, at the minimum, the following five vital, new realities of product and service creation that must be factored in, far before implementation. We have to make a transformation, and it has to start early in the conception process, in the very way we think through and develop our next products and services.

The New Realities of Product and Service Creation

1. Innovation is a discipline, driving nimbler responses to market changes, and a process, underlying every decision about product development.
2. The lines of demarcation between a pure product and a pure service are not sacred: products can expand into services (Best Buy's *Geek Squad*) and services can come bundled with products (iPad with apps preinstalled).
3. The canvas for distribution and competition is global.
4. Design matters.
5. The signs of our future already exist in adjacent and even unrelated industries, and provide insights into consumer preferences.

1. Innovation is a discipline, driving nimbler responses to market changes, and a process, underlying every decision about product development

Scott Wilson, product designer whose studio MINIMAL designed the Xbox 360 + Kinect Sensor, did a double whammy in the field of innovation. First, he designed cool watchbands for the iPod nano he called the TikTok + LunaTik, high-tech gadgets that double as a watch and a sound machine (think Dick Tracy). That was innovative enough. But Scott's second feat came from the world of funding. He had a problem: in order to even go into production, he had to have the money in hand just to fulfill the orders that were flying in the door. But, the traditional way of raising capital would have cost him too much time and slowed down his production schedule. Scott was facing a classic inventor's dilemma—how to go from orders to operations without losing his shirt.

Enter Kickstarter, an innovator in connecting creative ideas with funders. Speeding up the process of raising funds for marketable products, Kickstarter offers a novel outlet for acquiring capital that bypasses investment structures based on the traditional venture capital model. According to Scott, the Kickstarter model solved the classic chicken and egg syndrome usually associated with his inventions. "I knew that there must be a better way to gauge people's interest about TikTok + LunaTik. So I created a Kickstarter campaign that featured my invention and allowed the preorders to fund the production instead of having to raise capital from venture backers and waiting to see if orders would follow once we manufactured the initial production run" (Wilson, 2011).

Kickstarter put Scott's idea on the block and allowed people to crowd source support with small donations that were structured as start-up capital for the Nano Watch. Scott brought his design to the open market and, within a month, raised $1 million of grassroots funds (Thompson, 2009).

2. The lines of demarcation between a pure product and a pure service are not sacred: products can bleed into services and services can come bundled with products

I once worked with a pipeline marketing executive who was charged with driving double digit growth in the field of what's called pigging—cleaning pipelines with "intelligent pigs" that help protect against explosions. To increase revenues, he had been working with specialized chemists who had come up with a magic formula with the firepower to make the pigging process safer. Their newly-minted product, a pre-pigging surfactant that made the pipes more slippery, enabled the pigs to flow through with less friction. Coated with that "soap," the pigs could do a much better job of cleaning out the grit that built up in the pipelines and that led to blowups. It was a win-win: From the vantage point of the chemists, this innovative chemical formulation would position them as industry innovators; from the perspective of the customers, the formula would finally offer them a new defense against the hazards of transporting oil through their pipelines. He put the product's formulation in place and it was a success. The customer need for safety had already been established: safer pigging meant fewer explosions in the long run.

Dodge got smart. It bundled. It went to Mopar, its service arm, to make its own app for its customers. Available on iPhone, Blackberry, and Android phones, the app offers customer care, 24-hour roadside assistance, and a way for Dodge car owners to connect with one another. And, with the click of a button, Dodge drivers can find the manual, operating instructions, maintenance due dates, warranty information, and other important issues pertinent to their particular car purchase.

Somehow, though, initial sales were not very profitable, and even though the pipeline company had a better mousetrap, it didn't necessarily translate to more mice, or more customers. To the customer, the difference between a pig that was slippery and one that wasn't just wasn't worth much more money. They couldn't see the gain. So he was stuck with a promise for new profits and a product that couldn't gain traction.

Then the marketing executive got inspired. It wasn't until he applied insights from several different industries, marrying his pigging product with a service that *was* on the minds of his customers, that he discovered how wrapping his surfactant in something else was his route to long-term profitability. In other words, he bundled.

The idea came to him like this: On one of his visits to the pipeline's site, he ran into the company's risk officer who was up in arms about the news of a recent explosion. Even though the explosion wasn't related to the risk officer's company, it brought the inherent risk of the pipeline industry to the top of his agenda. The moment of eureka came when the risk officer asked out loud for a pipeline safety guarantee, albeit a bit tongue in cheek. There is no such thing as a safety guarantee in such an unpredictable and often dangerous industry. But the pipeline company-had exactly that, in a slightly different form.

Recognizing the opportunity, he moved quickly. To address the risk officer's concerns, he studied three services—the warranty programs of computer companies, risk management tables, and actuarial illustrations for insurance—and eventually designed a wrap-around service for cleaning pipelines that included their surfactant. The customers got what they wanted—less risk—and paid a premium price for both.

Adding services to products can reverse a downward spiral when profit margins are slim. But it can also add value to the products they come with. Think of a home security system married to a great monitoring service, or powerful enterprise software bundled with a 24/7 help desk and ongoing training.

Heartland Payment Systems: Fundamentally Changing the Strategic Competitive Game

Steve Elefant, chief information officer for Heartland Payment Systems, a credit card and online transaction processing firm that handles 5.5 billion

transactions a year, outlines how his company responded once it realized its electronic payment services were vulnerable to an unprecedented level of risk. Security breaches masterminded by 14-year-old
hackers—like the ones characterized in the movie *War Games*—were a
far cry from the realities Heartland faced. What Heartland stood up
against were extraordinarily sophisticated and well-organized experts
whose surreptitious actions succeeded in raising the level in the game of
protection.

Heartland was at a crossroads: protection against risk had become
more than a safety measure. The necessity for security had escalated
with greater technology and increased ease of intrusion. Security moved
up on the company's corporate radar, and its leadership invested time,
talent, resources, and whatever it would take to tackle this very large,
complex threat.

Elefant explains how the path that led Heartland toward security
was a means to an end of an entirely new suite of offerings that included
end-to-end solutions and forced Heartland to migrate from the transaction fee processing business to manufacturing, eventually promoting
itself to become both a leader—in setting systems standards with its
competitors—and a collaborator—using innovative strategies with government entities like the FBI and the United States Secret Service
(USSS). The stakes were high and Heartland had to race forward to meet
the challenge.

"We made a lot of significant strategic changes to put security front
and center. First, we used to buy payment terminals from third party
manufacturers. We discovered with this new shift in mindset about the
escalating risk in the area of security that the levels of security we could
get from the existing manufacturers weren't sufficient. So we now con-
tract-manufacture our own hardware hardwired with Tamper-Resistant
Security Modules that meet our standards. That way, we can guarantee
that the encryption of the data remains intact from the beginning to end
of the data processing. Second, we realized that if the networks of
experts in accessing data without authorization were well-organized, we
needed to lead the charge in organizing 'the good guys,' our peers who
faced the same threats. That was how Heartland led the charge to band

together a group of peers into the Payment Processors Information Sharing Council."

Heartland started as a service (payment processing) and, because of the security threat, added a product (it couldn't find any manufacturers to make a product up to their new level of specifications). With pigging, the pipeline company had a product (surfactant) and added a wraparound service to it (to make it more valuable to customers at first glance). It is the bundled nature of each (product with service OR service with product) that was the strength for both companies. The bundle solves the customer's true need and offers a turnkey, end-to-end solution.

Product-plus-service bundles can create a much richer relationship between an innovative product and the customers who use it.

3. The canvas for distribution and competition is global

Within the past decade, companies from P.F. Chang's, to FM Global, to Caterpillar have all shifted their focus to embrace global expansion and reap the rewards that were once upon a time only in the line of sight ofglobal giants like Johnson & Johnson and DHL. Even domestically geared companies have to quickly come up to speed on the forces of global economies, cultural priorities, consumer preferences, and market forces, sometimes winning big (as was the case in the P.F. Chang's entry into relationships with global partners in the Philippines, see the P.F. Chang's case study on page 145) and sometimes losing their shirts by misreading the market opportunity or running across some bad luck (like McDonald's experience in Iceland).

Domestic car manufacturers are not immune to this new reality: the world of business continues to grow and to make "local" more and more transcontinental. And it means a roller coaster of opportunities and challenges that will continue to put pressure on the auto industry for years to come. Global partnerships in the auto industry have set the pace for new opportunities, like the GM-SAIC collaboration in 2010 (World Expo Shanghai, 2010) that connected General Motors with the Chinese market. More than 2.2 million Chinese consumers lined up to see GM's EN-V, a revolutionary, newly designed "personal transporter." This is a good sign that innovation might offer GM leverage in the vast global opportunity in China. However, during the same year, China moved ahead of U.S. automakers as the number one manufacturer of light vehicles.

BYD, a Chinese car and battery maker, was ahead of its American and Japanese competitors with the first battery-operated hybrid car, mostly because it can make its batteries in-house. The batteries use the company's own lithium-ion ferrous phosphate battery, a solution that lasts longer and costs less to make. A car running on those batteries can go for 200 miles. So far, no other companies come close to the impact BYD is having on the world market, and the auto industry itself.

Global markets and collaborators have evolved into a world of global competitors, challenging the once-held dominance in the auto industry. And the auto industry is not alone.

Despite in-depth analysis and planning of an ambitious global expansion, in 2006, Wal-Mart exited from Korea and Germany. It discovered that its large box environment wasn't a good fit for the culture and it faced unanticipated competitive challenges of the market. Yet, many companies look at global expansion as a strategic next step, under scoring the urgency to learn lessons from others to avert risks and piece together a global vision that will stick. How has the global equation affected your industry, and what will you do differently to stay ahead of change?

4. Design matters

Who would have predicted that sometimes, if we build it, they *will* come? Because they only come if it's what they want. But how will we know what *that* is?

Every company can benefit from better design. Target raised the flag on design as an effective strategy to democratize designer-created goods, once only available as luxury items. It did so by inviting a team of artists, like Philippe Starck and Michael Graves, and foodies, like Giada de Laurentis, to create affordable but stylish lines of consumer goods and sales sky-rocketed. In its pharmacies, Target also made itself known with user-friendly design that used larger type and sported easier-to-open bottles. And Archer Farms' packaged goods built a new model in innovation with rounded edges on its cereal boxes and resealable snack bags. The innovator went from being just Target (with a hard-pronounced 'get') to the French "Tar-zhay" in the popular lexicon as the general public shifted to the mindset that good design was for everyone. Target had become more than a low-cost provider. A new standard was set, and design upgraded from nice-to-have to a crave-able driver of consumer demand.

Five years ago, corporate executives had bookshelves lined with titles that would have prepared them to fight a war or to make sure that their trains would run on time, collected as inspiration for how leaders should approach strategy and communicate it to employees. Process maps were the dominant graphic (to match the PowerPoint slides that dominated the board room presentations). Bookshelves now feature titles on design and data visualization like Edward Tufte's *Envisioning Information* (1990), Dan Roam's *The Back of a Napkin* (2008), Gestalten's *Data Flow* (Klanten, et al., 2008), and Alexander Osterwalder and Yves Pigneur's *Business Model Generation* (2010) to match the new approaches to presentations in the board rooms, featuring Pivot tables and Tableau.

The story of the Nano Watch, the record sales of the i-everything, the flurry of book readers, and hundreds of other beautifully-designed, elegantly-engineered products brought to market make the case that great design is a must-have for great products. But still, design can work against you. Take the Texas-based online window blind manufacturer that upgraded the aesthetics of its user interface and logo, only to discover customer traffic was significantly higher on the older, uglier site. The CEO said that in *his* industry, good design and user friendliness sometimes ran in opposite directions. Aesthetics did play a role in his overall user interface, but wouldn't win out over customer transactions when "uglier" design proved to drive more traffic.

That might be the exception to the rule; because, on the other hand, design is the killer app for many companies. Consider Coca Cola, a company with an ability to innovate that has long distinguished it as one of the leading brands in the world.

> Coke took stock of the new competitive forces it faced in 2008 and charged its newly assembled high-powered design team to breathe new life into its sales. It took design so seriously it transformed everything—from the design of refrigerated coolers to a reintroduction of aluminum bottles—to keep pace with changing tastes, stay ahead of the aesthetics of a crowded market, and insert innovation into operations, distribution, and marketing to drive a new surge of growth.

Though we can't depend on the power of design to solve every product development issue, we also can't ignore the power of visual communication in design elegance. Businesses today have to face the fact that visual sophistication has gone mainstream, starting with PowerPoint graphs in the '80s and '90s that visually communicated company goals. Today's leaders should all embrace the power of design and data visualization as new tools to envision and communicate their strategic vision.

There are other signs that design sense has become a mandatory component of strategic product planning. Companies like IDEO, Frog Design, Jump Associates, XPlane, and dozens of others with sophisticated design-based strategy processes work with companies like P&G, GE, Target, Qualcomm, Kaiser Permanente, and Steelcase to accelerate product creation (like Target's *Kitchen in a Box*, designed by Todd Oldham) that resonates with consumers (Segal, 2010).

Design has been democratized, so now our customers crave it. Design has been proven to clarify the explanations of our data, so now our presentations benefit. Design has played out as a driver of consumer demand, so now product innovation has become a new standard for all of us.

When companies invite "design thinking" into their mainstream, they can accelerate the pace of success.

5. The signs of our future already exist in adjacent and even unrelated industries

I sat in on a strategy session with the executives of a large shopping mall during the recent economic slump, and it was depressing—its real estate investments were upside-down, retail tenants were experiencing dismal sales, consumer spending had slowed to a near halt, no new tenants were sitting on the sidelines to rent square footage, and there were no signs of relief.

After the initial recitation of challenges, the tenor of the meeting changed completely when one participant started listing companies that were thriving in spite of the downturn: bankruptcy attorneys, online education providers, fast food restaurants, social service providers, eBay, i-everything apps, and pop-up shops (temporary retailers that rely on social media for word-of-mouth advertising).

Based on the inspiration from that list, the group transformed its discussion from hopeless products and services to trendy ones that would resonate with tenants and consumers. What was striking about

Kaboodle: The Mall Goes Online

With a mall that's viral and easily accessed on the Internet, what happens to every teenage girl's banter that we used to hear as she once passed each and every window display that caught her eye in real life? Hearst certainly thinks that innocent chatter can be saved, evidenced by its acquisition of Kaboodle, a social networking shopping site.

With Kaboodle, shoppers can upload pictures of stuff on Kaboodle's Web site that they're interested in buying, and get that crucial second opinion. It's no different than having an optician take a picture of you in your new glasses, so you can know for sure it's the right choice by asking for feedback from that person you trust.

With Kaboodle, Hearst moved up to an online readership, and reached more advertisers at the same time. And more customers— for both Hearst and retailers. Kaboodle boasts a population of 2.2 million visitors, mostly women with good taste. As owner of *Marie Claire* and *Cosmopolitan*, Hearst can draw a clear benefit from the collaboration.

the conversation was that all of the successful examples were from outside the shopping mall industry. The team had made the leap from a strategy that worked in one industry to one that could work in another—theirs.

The process transformed the obvious ideas first, like providing an option of temporary leases to seasonal shops like Toys "R" Us Holiday Express. But the mall managers also devised concepts like shopping apps, onsite health services, and entrepreneurism courses.

Every business needs to approach product development with a process that maps customer trends from outside its industry to its own. Make your own list of examples of great products from other industries that can inform your company's product development strategy and connect the dots to concepts that will drive growth.

Find Your Next Products and Services 🔍 find

1. Catch up to the pace of change with the whir of social media, new entrants into your company's industry, and new customer preferences.
2. Start the pipeline of innovative ideas flowing. Product innovation isn't a fad, but a competitive accelerator.
3. Keep an eye on your competition. Put yourself in your customer's shoes and look at your products from their vantage point. Where are they getting the most usefulness from your products? What are the alternatives?
4. Watch the basics like food, clothing, and shelter for clues on what customers will need no matter what. What can trends like fast-casual dining, clothing exchange Websites, and modular housing tell you about some of the basics that are changing for your products and your companies?
5. Watch for hot products. Monitor the popular gadgets to see how consumer tastes are evolving. Yesterday's ATMs became this year's mobile phone check scanning software. How can those new platforms catapult your products to a new level?

Chapter

5

Customer
Impact

Talk the Talk: The Art of Customership through Dynamic Conversation

Your Customer Impact Self-Diagnostic

1. Food, clothing, and shelter just don't cut it anymore when it comes to customers. The i-revolution has certainly proved that. So what is it that your customers need? Or want? And are you in touch with their deeper needs, or are you still busy making what you've always made, hoping they still want to buy it?
2. A more demanding customer isn't necessarily a bad thing. It also means a world of possibilities you haven't yet discovered for meeting those broader, ever-changing demands. How do you uncover these untapped opportunities?
3. Your customer will necessarily change, and has perhaps already changed, without your knowing it. Who will your next customer be?
4. The new world brought with it more options for customers, along with a forum with which customers can exchange advice, feedback, and product reviews. Have you adapted to the new customer dynamic of open product and service commentary?
5. Some companies seem to have figured it out. What exactly have they figured out and what can you learn from them?

The days of forecasting as a guarantee have passed. So have the days of labeling and identifying customers using simple data like demographic groupings. Back then—when the pace of change in the world of customer feedback wasn't as fast as the instant and massive customer response made possible by the Internet—the business environment moved at a much slower pace, customer loyalty was a given, and defection the exception, so that figuring out who our next customers would be was as simple as looking at our current ones. It was a time when brick-and-mortar stores dominated the commercial arena and live sales people cemented customer loyalty through great relationships. That meant that the people who bought what we were selling went *inside our store.* And though those retailers thought they could convert 100 percent of shoppers to a purchase, consumer research showed otherwise, that it was actually a rate of less than 50 percent (Underhill, 1999). Our perceptions of customer purchasing habits may have come a long way, but we're still mentally stuck in a time when that relationship with a knowledgeable, friendly clerk reigned supreme. That's all changing, and it has been. Even in that delicate transition from a three-dimensional building to a one-dimensional computer screen, we could count on customers to fall for us—and our brand—the minute they ordered and we gave them everything they needed: the product and a great purchasing experience.

Not any longer. Even the way we gauge customer satisfaction has shifted. Then we could assess our customership skills easily and reliably, in terms of product offerings and the customer experience, with company-controlled activities like focus groups, customer satisfaction surveys, net promoter scorecards, and other systematic tools designed to keep our companies safe from customer abandonment. We studied our competition, measured customer preference for us versus them, tweaked our offerings to stay ahead of the Joneses, and kept new competitors in our line of sight. We were taught classic rules for listening to customers, based on the Betty Crocker era of brilliant, focus-group-based insights that revealed the importance of a simple-to-make cake mix, but not so simple that the housewife didn't feel like she was still cooking. It was a time that led to decades of product innovations designed to lure customers to our wares. The rule of the day was "if you want to know what the consumer thinks, then listen" (Goebert and Rosenthal, 2002).

Placecast: The Internet Knows Where You've Been

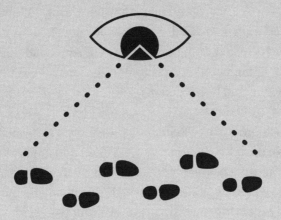

There's a new-generation way to capture customer input based on their online shopping patterns. Companies like Crimson Hexagon and Radian6 can aggregate data to answer questions like, "Did that person log onto Kayak first to compare pricing and then head to United.com to actually purchase his or her airline ticket, or did he or she log onto United.com first and then head to a discounter to purchase the ticket? " Take that skill set to the extreme, and it gets creepy—making us realize that companies can know almost too much about us. Viewed from a more positive perspective, we can envision a world in which companies can provide very "customerized" experiences and products. Alistair Goodman, CEO of Placecast, is passionate about personalization. He's even more passionate about privacy and the need to respect customers' personal data above all else. Placecast's software starts with an opt-in policy in which one customer can tell the company that she loves biking and her spouse can tell it that he loves to run. When the couple lands in Denver, the woman will receive notifications about bike paths and her husband will get maps of running trails. Data analytics is the next frontier in understanding customer preferences and brings a new dimension to the generation of customer research once fueled by focus groups and feedback surveys.

The company-centric ask-and-listen-based system won't work today; it's out of step with the new customer dynamic, in which consumers play a more active role in setting the agenda. We're playing a competitive game that catches one-time mavericks off-guard with every new entrant that steals the prime spot in customer affection. Businesses that can't tune into the new rules, like newspapers with eroding subscriber bases, run the risk of going extinct. Companies that have evolved, however, have been rewarded with increased customer loyalty. Take as an example Comcast, which created Comcast Cares as a forum to address customer complaints and increased its customer responsiveness, resulting in higher satisfaction and decreased defection (Gonzalez, 2011).

We're starting to question our long-held views about cause and effect with customers. We've been forced to abandon any belief that our businesses can push information to the market and automatically dominate the airwaves of customer perception. As business leaders today, we need a new rulebook to guide us through this chaotic era of customer psychology. The old dogmas like "the customer is always right" and "great customer service will lead to loyal, raving fans" no longer apply.

When Ken Blanchard's *Raving Fans* book came out in 1993, its powerful message to business leaders couldn't be ignored and guided years of customer-facing initiatives: "If you really want to 'own' a customer, if you want a booming business, you have to go beyond satisfied customers and create Raving Fans" (Blanchard and Bowles, 1993). But as the world of commerce heated up and sped up, we left the comfort of limited channels of customer feedback. In this new era, we can no longer apply the Betty Crocker rules for listening and the raving fans rules for service that will lead to customer loyalty.

We're left free-falling without a safety net. No longer can we do x for a customer response of y. That y might never come. Too often, we run the risk of being blindsided by the other guy, as illustrated by the example of MySpace and Facebook (Gaudin, 2010). MySpace led the race in social networking, but the other guy, Facebook, came along and did it better. According to Ezra Gottheil, an analyst with Technology Business Research, "My Space stopped being the place to find and connect to friends a long time ago." But the need was still there, so Facebook filled it.

The Facebook-MySpace story isn't just an example of what went wrong and why. It illustrates a need that has revolutionized customership.

The product and service discussion belongs to the customers now, and failure to correctly interpret the perception of our products in that forum can make the difference between life and death for our companies. How can we stay ahead of the customers in this new, dynamic customer landscape?

We all need to keep our eyes and ears open, and to see and hear what our customers are telling us in new ways, using new information. Our mindset needs to shift, from a brick-and-mortar focus-group mindset to a social networking one. We need to ratchet up our listening to be faster, better attuned to today's modes of communication, and more responsive than we were even a few years ago.

Customerization in today's market has very different rules. And no company is invincible from the rules of the new game. We live in a world where even gurus of customer impact like Starbucks can get it wrong, once taking the customers' shift toward the "foodie" culture to mean polenta would be a good add-on to coffee and pastries. (It wasn't.) In another example, reliance on consumer focus group feedback led a retirement community to design barracks for the 70-year-old parents of its 40-year-old customers (the children), except the septuagenarians didn't want living quarters, they wanted college dorms without curfews.

Traditional research methods based on the "You Asked for It, You Got It" Toyota mentality can't keep pace with the rapid dynamics in the new customer equation. Changing consumer preferences now cycle much too rapidly.

In 1922, long before the age of the supercomputer, Lewis Fry Richardson developed a highly accurate, manually produced weather forecasting model for the coming six-hour time period. The only problem was that it took six weeks to come up with a forecast. Think of your company in that same time crunch today: we need to read signals very quickly and respond nimbly to stay ahead. Our best next in these perplexing times is to follow the new rules on customers based on the new fundamentals driving customer responses and loyalty. We need to adapt to the new realities, look carefully at patterns from other industries, and graft those customer-specific elements of success onto our own companies.

The New Realities of Customer Impact

1. Customers don't tell us everything we need to know.
2. Innovation is shaping customer requirements at a faster pace than ever before.
3. The Internet has put the customer voice on speaker phone, and it's a party line. Companies risk becoming the victims of customer conversation unless they retool for the new dynamic.
4. Customer analytics have become an artful science of listening.
5. Global customer dynamics are on our doorstep.

1. Customers don't tell us everything we need to know

Eric Ryan, a style and branding expert, and Adam Lowry, a chemical engineer, couldn't compete with Procter and Gamble's Cheer and Tide, so they didn't even try. Instead, they came along and added up three elements of desire on the part of customers, which led to the creation of Method, a company dedicated to the once mundane world of household cleaners. But they did it their way. They asked different, better, questions of consumers. And in the end, they uncovered new information about what would drive consumer demand in soap, eventually building a company that just five years later was named by *Inc. Magazine* as the seventh fastest-growing business in the United States.

Eric Ryan's Five Rules of Branding the Cleaning Product Industry (Ryan, 2007)

1. Cleaning is a price-driven, low-interest commodity.
2. The experience doesn't matter.
3. Deep clean is the battle ground.
4. Eco-cleaners are a niche category.
5. You have to segment by product and be all things to all people.

A couple of decades ago, in the world of consumer research, products like Cheer and Tide got talked about in focus groups, the kind in which a dozen housewives sat around a table in a room with a one-way mirror.

For brands like that one the topic would be laundry, and the wives were impeccably moderated by a colleague whose script held questions like, "Which do you like more, brand A or brand B?" The women would dutifully spend the better part of an afternoon going through the drills flawlessly, and soon pleasant conversation would ensue about dirt, kids, laundry baskets, white clothes, colored clothes, and every other possible nuance of a task we all share—laundry. Many times, though, because of the structured nature of market research scripts during that time, the real question might never emerge: Do you really want to do laundry at all, or would you rather hire a cleaning service?

Fast forward to now when a new approach to the whole topic of laundry brewed in the minds of Lowry and Ryan, who were starting to wonder if there was an unspoken need that customers weren't able to articulate yet, a desire for a different type of soap that was more environmentally friendly and made for a better laundry experience.

Focus groups and round tables of the "typical" customer didn't last. And they didn't really work. Ryan and Lowry knew that and tried a different, more insightful tactic. How did they do it? They applied their experiences in marketing consumer goods for brands like the Gap, their observations about an emerging preference for environmentally friendly products, and the popularity of Target's new aesthetics in design to create a line of products that met the emerging customer demands in refreshing ways. They didn't assume that the path to the future was based on incremental improvements on what was already being done. Instead, Ryan and Lowry's team at Method saw signs of change and built novel solutions designed for maximum customer impact: loyalty to the Method brand came because Ryan and Lowry connected elements from the future that were already in the market in their line of environmentally friendly, aesthetically packaged soaps and household cleaners.

We haven't stopped measuring customer satisfaction. Such barometers have expanded to include tools like the net promoter score, credited with driving revenues for companies like Enterprise Rent-A-Car and Apple Retail Stores (Reichheld, 2006). And we've seen the rise of alternatives to the survey, the focus group, and the satisfaction score as experts in anthropology, psychology, and data mining have helped us interpret what our current and potential customers are trying to tell us (Kane, 1996).

However, in the customer-controlled, dynamic environment in which we need to read the future more rapidly than ever before, the first new imperative for tomorrow's businesses is this: Closely study your measures, metrics, and perceptions to keep the customers that you have, and make sure you're in touch with the new forces that now drive customer loyalty to get the customers that you want.

The new world isn't all about the advent of technology. It's also about intuition. And when the models and forecasts we've invested in tell us that our next product or service has to be A or B, we might be too prone to dismissing customer feedback that could lead us to C, D, or E, even if C, D, or E weren't options at first.

That means reading between the lines. We need to reconfigure the way we're learning about our customers, and how. Manipulating them to tell us what we want to hear doesn't translate to market leadership. Asking them for answers to the questions we need to hear, and building on those answers, can. For us it means we need to develop a sixth sense to hear what's not being said. We need to become more comfortable with ambiguity as we transition from a world of "either/or" to a world of "what-if" that truly resonates with our customers.

Do horses count? No. But you can get them to demonstrate knowledge of numbers by tricking them. But it means they're not really counting. A magician—our faithful master of illusion—could rig it for amazed viewers as a sideshow trick by leading a horse through a question-and-answer ritual designed to fool people that the horse canactually count. He'd ask the horse, "How much is two plus two?" and the horse would tap his foot four times in response. To many it looked like an example of an equestrian math whiz. Actually, it was all a show of a horse that had been trained to read the face of his trainer. Using the trainer's facial visual cues, the horse knew when to stop tapping. Put the trainer in another room and the horse did nothing. We can play that same trick with our customers but we're no better off. We're still a company making the wrong products for the right people, or the right products for the wrong people.

Are you one of them?

Are you seeing to your customers' unmet needs? Or are you starting with specific expectations about what your customers want then attempting to shoehorn them into wanting what you offer?

Are you prepared to add up your customer clues in new ways to build loyalty? Have you turned up the volume on the customer voice to tap into what they aren't saying they want, but do?

What signs can you see today of a new driver of customer impact that will make the difference for your company tomorrow?

2. Influences from other industries are shaping customer requirements at a faster pace than ever before

A couple of years ago, a large hotel chain faced a new threat: a new property in its backyard. The hotel's location—onsite at an airport—would not be able to be monopolized any longer. Wanting to maintain its high marks with customers, the hotel conducted an experiment to determine exactly what it would take to keep and maintain a loyal following. The checklist of things to mark off included all of the typical factors of guest satisfaction you'd expect: friendliness of staff at check-in, registration waiting lines, and room cleanliness. What threw off the hotel's higher-ups as they went through their questionnaire was an unexpected factor that made a very big difference to their guests: guests wanted to skip the check-in lines altogether. Hotel guests wondered why the driver of the airport shuttle couldn't simply look at their identification, call ahead to the hotel, and issue the paperwork and room key right there in the van or shuttle bus.

The model that the customers had in their minds wasn't based on the "inside" view of an hotelier, where the customer experience began at check-in. Instead, these customers based their expectations for service on their experiences with rental car companies whose Emerald Aisles and Gold Clubs had set a higher standard for responsiveness than the hotel queuing could ever match. The customers were in effect asking, "Why do I have to stand in line at all when you already have my reservation and payment information? Why can't the shuttle driver be our guest ambassador and phone in our registration information before we arrive, so we can head right to our rooms?"

Speed and Innovation in Hospitality

APA Hotel in Japan listened to what customers want—speed of processing—and began providing it. No longer must they wait for a live person to check them in, find their reservation, and process their room payment. Guests can now do it all themselves. And Matt Adams, from Hyatt Hotels and Resorts, followed suit, now keeping his eye on programs like Coolhunting.com to stay ahead of consumer tastes, like the rise of popularity in farmer's markets in urban cities. He applied his trend spotting to his hotel's food concessions, developing Market—a transformation of the hotel's New York lobby into a market-styled food concession for guests (Matt Adams, 2011).

In similar fashion, the managers of a sports franchise can look to hospitality and food service in how to best please its customers: the sports fan. Jamey Rootes, president of the Houston Texans football team, looks at the live experience:

"As President of the Houston Texans, I'm constantly working with our leadership team to make sure that the fan experience is at the absolute peak of what our organization can deliver. Our business leaders play our own versions of offense and defense when we design and deliver the systems and processes that create an over-the-top experience. When I say we play 'offense,' that means we ask ourselves, 'What are the kinds of things you can't do on your couch that we can take to the maximum level of engagement and enjoyment?' We look at tailgating, the sense of bonding and excitement of being part of an enthusiastic crowd, and the entertainment factors that enhance the game, everything from cheerleaders to music to traditions to special promotions. We build on everything that can give the fans an emotional lift and immerse ourselves in every aspect of that experience, learning and borrowing from the best in the business, whether that's from a concierge's hospitality at an upscale resort or what goes into the fun factor of Disney.

On the other hand, we also play 'defense,' which translates to examining absolutely every component of the fan experience from ordering tickets to traffic and parking to sitting in the stands. We challenge ourselves to win wherever we can, eliminating any potential negatives like long lines at concessions stands or traffic jams on the way to or from the game. When we wanted to make improvements in our in-seat food and beverage service, for example, we sent our managers out to learn from the McDonald's drive through.

I've learned that outside our industry there are people doing things that might look different than what we do, but are fundamentally the same as what we're trying to achieve—and they might be doing a superior job in areas we'd never imagine could move us to a higher level of execution or impact. Our friends at Starbucks say, 'Don't let the coffee down'—the coffee may be great, but people love the experience inside of a Starbucks. That's exactly what we're going for. People love the game of football and we can't let the game down by not adding to the experience with a huge boost of spirit, energy, and vision."

Perry Farrell, legendary rock frontman for Jane's Addiction and founder of Lollapalooza, says, "When I think about 'nexts' in the world of music, I always look beyond the strategies of other musicians for inspiration. I start with a thought about fans—what does a live event provide for fans that fills the seats? What will capture the enthusiasm of the crowds? Years ago, that question led me to look at how sporting events attracted crowds ... eventually giving birth to Lollapolooza, a music festival that offered a new setting for experiencing live music. Whether its lessons learned from NASCAR or the NFL, there's a lot to master about how to bring the thrill of live events to new levels."

Hotels, football teams, and rock music promoters can all look at how to attend to consumer preferences in not so obvious ways. Companies can even use internal staff dialogue to fine-tune the customer experience by getting ideas from their own talent.

Best Buy: Looking Within

Steve Bendt and Gary Koelling turned their focus inward instead of outward when it came to idea mining for their employer, Best Buy. Taking advantage of the Best Buy community, the two ad men put together something called Blue Shirt Nation, visiting stores all over the nation, and engaging in one-on-one conversation with Best Buy employees. This gave them a sense of what might work, and how. They then created an online open forum that let Best Buy staffers get together regularly to swap ideas and interact. People like to talk, especially when other people listen.

Best Buy's Blue Shirt Nation, established in 2006, was an early player in real-time, out-in-the-field customer insights. Best Buy set up a site that started out like a big, electronic suggestion box, and evolved into an entire ecosystem of ideas, conversations, and relationships. The Blue Shirt Nation is credited with everything from redesigning in-store kiosks to completely rethinking its electronic game sales, all based on the strength of its electronic community.

Sometimes our customer's scorecard for us changes while we're looking somewhere else. Are you watching for new standards set by companies outside your company's industry in order to keep your company's competitive edge?

3. The Internet has put the customer voice on speakerphone, and it's a party line

Certainly the most dramatic shift facing businesses is the zoom into the era of electronic commerce and social media. Over the past five years, customers have flocked to the Internet in record numbers, and in the process, turned up the volume on their collective voices. What's emerging is a new world order for commerce that encompasses not only consumer goods and services but also new rules for business-to-business interactions, in which transparency and speed are the hallmarks of electronic transactions. Consumer-driven standards have all but replaced company-driven ones. In this new consumer world, online holiday spending is growing at a faster pace than in-store purchasing. Social media has become a powerful player, with companies like Pepsi shifting ad dollars from television to social media (DreamGrow Social Media n.d.). As a result, the rate of acceleration into a new customer landscape has increased to the point that we now find ourselves in an entirely new world of commerce. In the past hundred years, the definition of word-of-mouth advertising has been revolutionized. Key customers once spread catalogues to twelve of their friends (like Sears' 1905 "Iowazation" project), but now retailers port their wares to mobile devices and many have dropped paper altogether. Customers talk to each other publicly. Our companies' conversations are more exposed. Business transactions are more transparent than ever before. And it isn't just in retail. In the business-to-business environment, the online marketplace has given birth to reverse auctions, in whichcompanies put projects out to bid to multiple vendors. This inverts the forces in the bidding equation; the new customer requirement can be placed front and center in the auction, and multiple potential vendors from anywhere in the world can jump through the hoops to prove they can win the bid.

One of the most dramatic challenges we now face is having reached the end of the road for the "information push." Since the advent of chat rooms and social media, businesses have to grapple with the fact that they can't simply push information to the customer without having them push back, talk back, and discuss it amongst themselves. Consumer-to-Consumer (C2C) conversations are major influences in our marketplace. And, in the latest wrinkle in Consumer-to-Business (C2B) communication, customers talk about us to each other, outside the confines of our walls, and then write back to us at a record pace. It's as if they notice our every move: "Why did you take chicken fingers off the menu at Houston's?," "Why is the live chat queue so long at Home Depot today?," or "Why did the United Airlines close down its customer service line during the snow storm?"

Conventional wisdom for wowing customers no longer applies. The new skill that will drive customer impact is the mastery of a new world where platforms like Twitter invite everyone to "Join the Conversation." In today's world of Facebook, blogging, and recommendation sites, the old rules of telling customers why they should love us have been replaced with a dynamic in which the customers voice their opinions about us openly and expect to engage in a dialogue. Whether it's a travel site like Trip Advisor, on which customer comments fuel hotel and travel ratings, or Connected Moms, on which women share insights about baby products, the customer is engaged in open discussions about what companies offer today and which products get customer thumbs up recommendations or not.

Follow these new rules for harnessing the power of speed, transparency, access, and multiple-way conversations that will provide competitive advantage to all of our companies in the coming decade.

The Rules of Competitive Customership

1. Open up the floodgates of input from your employees, starting gradually, and listen to ideas for improving your business based on the observations of the people in the field (Haugen, 2009).
2. Forget controlling the conversation. Thin out the firewall between your company and your customers. Whether in a consumer field, a service industry, or a business-to-business environment, you can host dialogues and curate content about your industry.

3. Build full-out social media sites that engage your customers.
4. Capture the movement of the crowd. Watch for what wows your customers with tools like Empathica, net promoter scores, and customer relationship management (CRM) software (for example, Goldmine and Sales force). Test your ideas on a small scale with some of your fans, advocates, and current and prospective customers. Look for signs of traction that could build to a new offering.

4. Customer analytics have become an artful science of listening

More technology isn't always better. Customer relationship management software, while powerful, can send companies down a rabbit hole of rapid meaningless information tracking. We need to be careful about getting too excited about technology for its own sake: we need to think before we measure.

One computer company I worked with did just that. By tuning into the patterns of some key analytics, that company was, in fact, able to separate noise from meaning, leading itself to a competitive advantage in reducing annoying errors (modems that didn't work out of the box) and anticipating customer requirements (adding live chat to their help desk's online offerings).

However, such cautionary insight holds true for all business leaders trying to respond to the growing ecosystem of trackable data of customer preferences. In this new age of online relationships, it can make a difference in corporate growth. But we still need to exercise restraint in how far we go with all of this information.

A CRM 101 course would include a walkthrough of the wizardry of analytics, and all the options now available to us, from free Google and Yahoo services to a host of for-a-charge solutions that can decipher customers' preferences using sophisticated, analytical engines. And yes, it is important to exercise the new muscles of online customer dynamics; and yes, it is important to read the data with the customer's point of view front and center. However, updating to the customer relationship management 2.0 version can be inspired by companies who have created new mechanisms for driving customer impact, even if they aren't successful at impacting anyone or anything but numbers.

Procter & Gamble: Bringing Up Baby

Procter & Gamble transformed the conversation about Pampers diapers from ideal shelf placement, couponing, and direct mail to staying connected with "women who are pregnant or thinking about becoming pregnant and [staying] involved with [those women's] lives with a wide range of products and services throughout their first few years of the baby's life" (Rae, 2009). Journeys like that begin with an analysis of information about who's pregnant and thinking of becoming pregnant through conversations that are publically exposed in inquiries about related products, blog entries, and fans of spokespersons on related topics.

The year 2010's trend lists underscored the rising importance of the new gold standards: (1) online activities and their effect on offline sales are key to customer loyalty, and (2) customer service and interaction with your company is now social (DreamGrow Social Media n.d.). But, taking a proactive, competitive approach means that business leaders have to read signals in new ways.

A good starting point for designing Web analytics includes hypothesis testing to setting up assumptions and tracking results, and the development of tag clouds and keyword trees to understand the patterns of the searches that are driving your customers and potential customers to your field of interest. For Pampers, those analytics would include searches about multiple topics in categories related to babies, young children, work and life balance, and health. The patterns of what you see can also reveal hidden opportunities. For example, if a hotel tracks a guest's search for spas after the guest books a room, the hotel can understand the mindset, or psychographics, of the traveler, and develop add-on services like the W Hotel's partnership with Bliss Spas.

Transform your thinking from a mental picture of a spreadsheet of customer-defined categories into a sticky planet in a solar system of customer data. Be prepared to come up with ideas through conversations that are occurring close by (motherhood), then zoom into your own company's "planet" (diapers), and locate the universe where your products reside in your customers' worlds.

The downside of measurement is that it can paralyze decision making. Having a narrow view of what we're looking for can help us zero in on answers we can act on.

Luckily, the speed of e-commerce allows a business to test hunches and retool quickly.

5. Global customer dynamics are on our doorstep

In 2007, *South Florida CEO Magazine* reported it was time for the local companies in the Miami region to "go global," citing dozens of Chinese-owned companies like Envision Peripherals, Latin American enterprises like Brazilian-headquartered Embraer, and European companies like Nokia that had moved significant United States – based operations to within a stone's throw of the Miami International Airport. According to the article, the global companies had already brought $26 billion and 230,000 people to the region (Hemlock and Cohen, 2007). The flip side of that positive economic report is that the presence of global competition, whether in our backyards or otherwise, puts every company on the global stage.

The positive side of an era when global commerce is the rule and not the exception is that opportunities present themselves through new coincidences. One of the biggest surprises to P.F. Chang's was the number of foreign partners who entered the equation as possible suitors for licensing agreements in the Middle East, Mexico, and the Philippines, and met the companywith a nod of familiarity as they started negotiations, "Oh yes, I ate at P.F. Chang's all the time when I was at university in the United States" (Welborn, 2011). That recognition accelerated the building of the relationship and shortened the learning curve required to become a valuable licensing partner. Customer relationships can be cemented on a global basis through partnerships and collaborations.

Customer impact depends on relationships, and in the landscape where relationships with foreign competitors are much easier to forge either because of proximity (like in the cases of headquarters moving to the United States), familiarity (because companies like BP and Unilever have such a large domestic footprint), or serendipity (like P.F. Chang's experience with potential licensees), achieving customer impact with a global perspective is a new reality of the game.

Find Your Next in Customer Impact   🔍 find

1. Turn up the volume on the voice of your current customers. Equip employees with tools to view your company from a customer's point of view.
2. Teach your teams to have a sixth sense for untapped customer opportunities. Has the bar been getting higher in a key customer preference in another industry that you can build into your offerings?
3. Tap into the power of the online conversation. Transform your information push approach into an open forum for ideas and input.
4. Watch the edges of innovation in the customer arena. Track the technologies and products that are turning heads today that you can inject into your plans for the future.
5. Look inside your industry and look across industries, including to global companies that you didn't realize were in your customers' line of sight. Who is capturing the attention of your customers today?

Chapter
6

Talent, Leadership, and Culture

Leadership 2.0: The New Right Stuff for People and Culture

Your Talent and Leadership Self-Diagnostic

1. What fresh models for leadership keep pace with the today's new business realities?
2. What new rules are transforming the people in our organizations into a motivated workforce?
3. What rewards must we embed in our organizations? How do we create incentives that align with our priorities?
4. How can our company's culture create a great place to work and transform our company into a market leader?
5. How can we exercise the new corporate muscles that will propel us forward in today's global, technology-enabled, diverse, competitive landscape?

Even military leaders are rethinking their assumptions. At Technology Engineering Design (TED) in March 2011, Stanley McChrystal talked about his evolution from army lieutenant to four-star general, and what it means to be a leader in the military. He faced a more complex, less-predictable environment, because the organization of combat had outgrown the classic, historic battle planning that used football-play look-alike diagrams and logistics planning formulas. In today's wars, he explained, both women and men between the ages of 18 and 48 now readily engage in combat, multiple complex technologies are deployed to infiltrate enemy lines, and the command-and-control structure is highly irrelevant and ineffective.

Then, he told a story of a personally defining moment, September 11, 2001, when he came to question what all of his West Point training and combat experience had taught him about the role of a leader and he had to recalibrate what it would take to stay "credible and legitimate" as the commander of his troops. He was preparing for a military intervention in Afghanistan, and addressed the soldiers in what he believed would be a moment of shared purpose and inspiration before they headed off to fight. For McChrystal, that shared sense of mission for all of the troops began with September 11, when the United States felt vulnerable to foreign forces and the population was motivated to protect the safety of its country. But that doesn't mean his troops would remember it as he did.

He started the talk with a question he had used for years as a means to focus the team toward a unified purpose. "Where were you on September 11?" Singling out one soldier in his line of sight, he asked it directly, "Where were you on nine-one-one?"

"Sir, I was in the sixth grade."

McChrystal had not been in the sixth grade. He was born in 1954. But his troops weren't. That was the four-star general's wake up call. That fateful day in September didn't impact them in the same way it had impacted him. The men he aimed to lead and their leader were not one and the same. On that day and over the course of many that would follow, McChrystal came to realize things had changed—the leadership philosophy that would make him relevant today wasn't what would have made him relevant yesterday. It needing updating and rehauling, or he'd lose his followers and their power. It was a new world of warcraft and not "his father's Oldsmobile."

We run companies, not military units. Still, despite our differences, all of us can learn from the process General McChrystal crafted as he set out to build a more responsive "leadership 2.0" model. He began by observing the factors that worked to motivate his people, coining a new term—"the inversion of expertise"—that captured his insight that his soldiers had values, modes of communication, and beliefs that were foreign to him. It meant he could learn just as much from the people in his organization as they could learn from him, without undermining their respect for him as a leader.

McChrystal created the new role of "reverse mentor," in which his colleagues and direct reports transformed from student to teacher, imparting essential new skills to peers and superiors. Reverse mentors were responsible for creating rules of the new Army, incorporating two of the driving forces that motivated the current-day soldiers—increased levels of transparency and responsiveness—into the day-to-day culture. And, over time, General McChrystal made himself into a new kind of leader that could bring out the best in his troops. For the general to be effective as a leader in this new environment he had to adapt and devise new strategies for connecting with and motivating the talent in his organization. That meant great change for the general. He learned their language, sending text messages to troops in remote locations and engaging in open dialogue about ethics of combat in civilian neighborhoods. Learning from his troops brought strength to his role as leader.

All leaders, whether operating in the currency of war or the currency of commerce, now face perplexing times. Business leaders are not immune to the realities the general experienced—those of an increasingly diverse workforce and a rapidly evolving work environment. We know the importance of attracting and retaining the people who bring our organizations to life, but when we analyze the messages and models that molded our visions of great leadership, we find we need to refit our leadership chassis with new parts that are designed to drive our organizations forward and fire up our people. For all the emphasis we place on innovative products and efficient processes in the world of business strategy, it turns out that people are the secret weapon.

Talent within an organization creates our company's point of difference, instills loyalty in our customers, and brings commitment, discovery, invention, novel ideas, and passion to what we do. In every great process, whether the Zappos customer experience, the software at SAS, the processes at Wegmans Food Markets (McGregor, 2010), or the call center innovations at L.L. Bean (McGregor, 2010), it's the employees who make the difference between an "also ran" and a "best of breed" company.

Yet, when it comes to actually figuring out how to attract and keep great people, even the most seasoned business leaders are challenged. The forces have shifted significantly, and our old bag of tricks no longer works: giving badges for years of service mixed with a year-end bonus isn't enough to keep anyone motivated. Top-down inspiration that was once the hallmark of our classic business role models has been replaced by a new tone of personal communication from boss to employee, like the one Tony Hsieh used in his classic memo to his employees at Zappos, following Amazon's $800 million investment in the spirited shoe startup. This is a perfect moment to reflect on the impact we're trying to assert in the relationships we create—with our organizations, customers, and partners—and rethink what we're striving to accomplish as we lead our companies toward our nexts.

A new tone for leadership communication by Tony Hsieh and a strong endorsement of culture as a value driver for an organization (Frommer, 2009)

Date: Wednesday, July 22, 2009
From: Tony Hsieh (CEO – Zappos.com)
To: All Zappos Employees
Subject: Zappos and Amazon

Today is a big day in Zappos history.

Over the next few days, you will probably read headlines that say "Amazon acquires Zappos" or "Zappos sells to Amazon." While those headlines are technically correct, they don't really properly convey the spirit of the transaction. (I personally would prefer the headline "Zappos and Amazon sitting in a tree ...")

Our culture at Zappos is unique and always evolving and changing, because one of our core values is to Embrace and Drive Change. What happens to our culture is up to us, which has always been true. Just like before, we are in control of our destiny and how our culture evolves.

A big part of the reason why Amazon is interested in us is because they recognize the value of our culture, our people, and our brand. Their desire is for us to continue to grow and develop our culture (and perhaps even a little bit of our culture may rub off on them).

First, I want to apologize for the suddenness of this announcement. As you know, one of our core values is to Build Open and Honest Relationships with Communication, and if I could have it my way, I would have shared much earlier that we were in discussions with Amazon so that all employees could be involved in the decision process that we went through along the way. Unfortunately, because Amazon is a public company, there are securities laws that prevented us from talking about this to most of our employees until today.

We learned that they truly wanted us to continue to build the Zappos brand and continue to build the Zappos culture in our own unique way. I think "unique" was their way of saying "fun and a little weird." :)

Amazon focuses on low prices, vast selection and convenience to make their customers happy, while Zappos does it through developing relationships, creating personal emotional connections, and delivering high touch ("WOW") customer service.

We realized that Amazon's resources, technology, and operational experience had the potential to greatly accelerate our growth so that we could grow the Zappos brand and culture even faster. On the flip side, through the process Amazon realized that it really was the case that our culture is the platform that enables us to deliver the Zappos experience to our customers. Jeff Bezos (CEO of Amazon) made it clear that he had a great deal of respect for our culture and that Amazon would look to protect it.

The Zappos brand will continue to be separate from the Amazon brand. Although we'll have access to many of Amazon's resources, we need to continue to build our brand and our culture just as we always have. Our mission remains the same: delivering happiness to all of our stakeholders, including our employees, our customers, and our vendors. (As a side note, we plan to continue to maintain the relationships that we have with our vendors ourselves, and Amazon will continue to maintain the relationships that they have with their vendors.)

... It's definitely an emotional day for me. The feelings I'm experiencing are similar to what I felt in college on graduation day: excitement about the future mixed with fond memories of the past. The last 10 years were an incredible ride, and I'm excited about what we will accomplish together over the next 10 years as we continue to grow Zappos!

Tony Hsieh
CEO – Zappos.com

Several new, fundamental realities exist that leaders have to integrate into their definition of leadership in the new world of commerce. We need to abandon some of the models we've built. We need to listen in new ways to the people we interact with. We need to form a new consciousness around what is now.

The New Realities of Talent and Leadership

1. Diversity is on the rise.
2. The onslaught of information is multiplying at speeds that make it impossible to keep up.
3. Employees can no longer be motivated by traditional reward systems.
4. The definition of leadership is changing.
5. Connections via technology demand the mastery of new communication skills.

1. Diversity is on the rise

The diversity of today's business landscape has created an ecosystem in which top-down leadership has become largely irrelevant. Baby boomers work with Gen Y-ers, and global collaborations are a common component of the new business equation. Even companies that don't have global operations are affected by the global trends in talent, product development, sourcing, technology, and economics. Some companies like Yum Brands have moved aggressively into the global arena, marked by its slogan, "Feed the World." People's cultural backgrounds follow them to the workplace and the best companies have learned how to integrate these diverse points of view into the fabric of their organizations.

Jodie L. Jiles, a seasoned community and corporate leader, explains his perspective on building a competitive work force with an eye toward the future. His advice adds a new sense of urgency to the call for diversity in leadership.

"Every organization's work force should mirror its customers and constituents—not just the ones we interact with today, but the ones who will be on our radar two to ten years from now." Jiles looks at the new realities facing virtually every company he interacts with and emphasizes the power of the mirror to teach every leader how to appeal to Millennials and Boomers at the same time, hire the best global talent, and expand our own horizons to anticipate the evolution of our customer base. A move toward a more diverse leadership pool has the power to insert tomorrow's consumer preferences into your company right now.

2. The onslaught of information is multiplying at speeds that make it impossible to keep up

Business is facing an "info-lanche," and the rate of new data infusion is too fast for any one individual or even any one team to manage. Organizations need new processes for tracking and integrating information, and leaders need to focus on enhancing their organization's responsiveness to a rapidly evolving set of facts and insights. Consider one simple example: customer satisfaction call monitoring. It's estimated that just in one aspect of a company's organization—call center management—data can be provided on every interaction of every call, giving those companies the opportunity to coach employees on instilling loyalty in their customers. That translates to a company like eLoyalty tracking 500 million hours of phone calls in a database, all ripe for analysis (Singel, 2011). Multiply that amount of data by all the processes and metrics aimed at keeping track of facts and it becomes clear just how challenging a feat that is, and near impossible to accomplish.

3. Employees can no longer be motivated only by traditional reward systems

Why does Teach for America attract such great young people despite its modest salaries? Because organizational rewards don't always match employee motivations. We can't avoid it. Employee turnover is a fact of life for companies. According to the Bureau of Labor Statistics, the baby boomers held approximately eleven jobs in their lifetimes, but early signs of the Gen Y and younger generations hint at a more migratory workforce. The rewards from yesterday, like the gold watches for years of service, don't stand a chance with such a mobile work force.

New thinking like that of Daniel Pink, author of *Drive, The Surprising Truth About What Motivates Us* (2009), points to new motivators in today's workplace, where it's not just bonuses that will incentivize employees, but many other proverbial carrots, including what Pink describes as an innate need to be self-directed, to learn, to create, and a "yearning to be part of something larger than ourselves."

Leaders today must create their companies futures. Talent 2.0 means highly engaged people bring our companies to life. You know that your competitive edge depends on attracting great people and keeping them revved up, engaged, focused, and on track. But how can you crack the code on what will feed them? And how can you set a course for the people in your company that strengthens your company's competitive edge?

4. The definition of leadership is changing

Leaders are living in a new world, but with tools—and mindsets, standards, and perceptions—that are out-of-date and no longer work. For example, leaders might want their organizations to be more innovative, but default to filling top positions with non-innovators. They might want to embrace diversity, but obey long-held biases that keep their employee base homogenous. According to Sharon Richmond, director of the Change Leadership Center of Excellence at Cisco Systems, it's time to refresh some of the assumptions and stereotypes about who can lead successfully, and retool our senior teams on integrating these new perspectives in the way they select and develop their leaders of the future (Richmond, 2010). "Leaders have to courageously reevaluate their own assumptions and mindsets, and specify the new leadership behaviors that will take them to their new next. Only then will their organizations be ready to describe and reward the behaviors needed to create and reach that future." At Cisco, the company's leadership model calls out "disrupt" as a critical leadership competencyand specifies that leaders should advance the business through constructive change, by actively considering alternative strategies, revenue models, and operating frameworks. Managers at all levels are evaluated based on this capability, and success in this area is important for career advancement.

5. Connections via technology demand the mastery of new communications skills

Technology has taken us from live to remote to virtual. Teams are routinely connected by e-mail, teleconferences, webinars, and telepresence. On the one hand, leaders need to learn how to provide remote inspiration and rework their communication so that it's effective in a newly "wired" one-to-many mode, but on the other hand, using those same technological tools, leaders can plug in to the wisdom, insights, ideas, creativity, spontaneity, and other input from the organization's workforce to stay ahead of trends, understand front-line perspectives, and read their competitive environment from a richer perspective than ever before. Technologies like Skype and Cisco's umi home telepresence service are ironing out the kinks in connecting users seamlessly between home and office (Corner, 2011). Leaders have to master these new skills to keep the spark in their people alive.

But, there is a footnote: leaders have to be conscious of the cultures they create and exercise a different set of muscles to lead successful innovation initiatives. The same rules that apply to building a brand apply to building a culture. The rise in importance of values like authenticity, transparency, and community input has spread from the customer to the employee. Now it's more than just looking the part, companies have to live it as well.

Leaders who invest in building a specific culture that supports a new competitive direction see the impact in very tangible ways, says Dave Scholotterback, chairman and CEO of CareFusion: "Corporate culture can be the single most powerful determinant of success or failure of any new initiative, whether it is a new direction for an existing company or an acquisition. You can't simply inoculate 3,000 employees once and assume [that the new habits required for success] are embedded for life" (Schlotterbeck, 2011).

The key is to start a process to determine the kind of future you see ahead of you, to think of those people who will help the organization realize its greatest potential, and to consciously design a culture that matches your vision; or do what Rebecca Campbell, Human Resources

Executive for JPMorgan Chase, describes as leading through change. Rebecca, who has traveled the globe preparing leaders for the future, makes it clear that the world of commerce is shifting so rapidly that waiting for things to calm down isn't a real option. She emphasizes the importance of clarity of purpose, something that can weather a wide range of market conditions and competitive dynamics (Campbell, 2011).

Jones Lang LaSalle: Leadership Development with Innovation

Jones Lang LaSalle (JLL) is a global real estate services and investment management firm with more than 36,000 employees with expertise in corporate solutions, energy and sustainability, tenant representation, and a wide range of related services. Emily Watkins, senior vice president for innovation and product development, views innovation as a critical component of competitive advantage, based on the viewpoint that an innovative culture drives better solutions and more powerful offerings for her company's clients worldwide.

With a suite of tools designed specifically to cast a wider net for idea generation, Watkins and her colleagues stimulate new thinking that feeds into JLL's product development process.

Here's one example of a recent experience that was significantly enhanced by the innovation tools: "We recently tackled the issue of staff recruitment in Latin America and you could see the impact of the innovation methodology. The traditional problem-solving approach would have started with our current operating guidelines as our point of departure and the result would have been improvements on that process. By contrast, with the innovation framework as our point of departure, the leaders in the group defined the problem differently and asked themselves a non-traditional question, 'Which organization does a phenomenal job of attracting, recruiting, and training top performers?' That led to an exploration of the Brazilian football league system and fast-forwarded the group to a whole new world of possibilities for how to shape our talent recruitment."

Scott Parazynski, chief medical officer and chief technology officer with The Methodist Hospital Research Institute (and former NASA astronaut and Everest summit team member), despite his own significant level of achievement, echoes the sentiment that a leader can create a culture that involves everyone in a shared sense of purpose. "There are many parallels that can be drawn between business and NASA, or challenging oneself on a Himalayan peak: enormous risk is involved, and failure *is* an option for those who are ill-prepared. Intensive study of the environment, knowing the strengths and weaknesses of every member of the team (including oneself), and being prepared for the absolute worst gives you confidence to truly excel under most circumstances. If adversity does strike, you'll also have the wherewithal to handle it for the best possible outcome. People often regard me as a risk taker or daredevil, but I'm most certainly neither. I work hard to prepare myself—and my team—for success, by understanding the risks and then actively managing them" (Parazynski, 2011).

Bob Johansen, author of *Leaders Make the Future: Ten New Leadership Skills For An Uncertain World*, takes the theme one step further: He believes the key to leading an organization is focusing on the skills that equip you for an uncertain future, like "immersive learning" and "dilemma flipping," so that you're able to see the future, and adapt more quickly than your competition (Johansen, 2009).

You know it when you see it (or don't). But that doesn't help with the how. I saw it when I first toured the offices of ScoreBig. That company does it with layout and design. You know what they do for a living as soon as you walk in the door of their office. The layout of the headquarters tells you everything you need to know: sports jerseys hang from the rafters, the office space of open seating has pods of collaborators working through highly stimulating projects, a palpable buzz hums in the air, and a break room of diversions ranges from football to ping pong. Sure enough, when the staff gathered later that same day for an employee meeting, the addition of five new jerseys to the symbolic collection that rested from the exposed pipes was on the agenda. The two co-founders gathered the team of fifty or so employees for a very important ceremony—honoring years of service with something more fitting than a gold watch.

In this case, ScoreBig's ceremony was a perfect embodiment of the company's unique culture—the employees celebrated (with hoots and hollers) the one-year anniversary of those five employees, and marked each announcement with a custom-designed shirt that reflected their number, every staff member had one, (numbers 9–14, in this case), and favorite sports team (cricket, soccer, football, tennis, and basketball). The co-founders' speeches were full of stories of long nights and personal insights that clearly demonstrated the value the startup placed on every single individual's contribution to the magic created inside its four walls.

A completely different, supposedly "people friendly" office of a large, global financial services firm in New York I once met with showed an opposite set of priorities. Its HR department had hung signs containing inspiring slogans that implied that the company was all about its customers, yet the artwork of the office—pictures of rocking chairs and still-life paintings—retained an air of corporate neutrality and staleness. No personal items were displayed on desks. In fact, there was a rule against it. It was an environment that posted words that said "people matter," in an atmosphere that communicated that they didn't.

Every company doesn't need to create a dormitory feel to keep young people comfortable; nor are large, multinational companies incapable of telegraphing their values in a meaningful way. But a culture that reinforces its leaders' visions and embodies the values and priorities they live by will be much more successful in attracting and retaining employees.

A newly hired CEO of a large multisite company I once met wanted to communicate his commitment to an open door policy so strongly that he actually made his first symbolic act the removal of the high walls around his office, followed by weekly fireside chat updates to the entire organization. Everyone picked up on the fact that the new leader valued dialogue with employees, and, over time, that one ritual set the tone for greater transparency and openness throughout the entire organization (and perhaps was responsible for its double-digit growth in revenues).

The key to culture is to look for signs and signals that transmit meaning about what your organization's priorities are, and disconnects that could erode progress toward a future vision in which you're saying one thing but your policies, behaviors, modes of communication, and other "intangibles" are saying something completely different.

A more proactive approach to culture can serve as a powerful reinforcement of a corporate strategic vision. Like the knocking down of a CEO's wall or the hanging of a sports jersey, small things can have a huge impact. By looking at our workplaces as an anthropologist or archeologist would, we can embed our offices with artifacts that support our beliefs and show our people who we are.

Picture the offices of an online brokerage firm I once worked with based in the Midwest that had acquired a smaller firm in New Jersey, but the acquiring firm hadn't done a great job of combining the two cultures. The head of the New Jersey office was dressed in a golf shirt on a "Casual Friday." While he spoke with words about how well the new parent company had integrated the subsidiary into the fold, his shirt, of course, still sported the logo from his old startup. People have trouble letting go. He certainly isn't the first and he won't be the last to hold on to what was. The verdict: the cultural cue told anyone paying attention that the acquisition had not fully taken hold and that there was work to be done before New Jersey felt the love from the Midwest.

If you played the role of stranger today and walked through your own workplace with fresh eyes, what would the signs and signals of your culture communicate to you?

Best is not one size fits all; the leadership adage that derives from that insight: "If you plant peaches, do not look for pears."

It makes sense that if people see your company as a great place to work, they'll be motivated to be a great person to work *with*, which makes measurements like "Best Place to Work" very significant in tracking how well your business is doing in the people part of the competitive equation. Traditionally, *Fortune* magazine would release its rankings for "Best Places to Work" and readers might come to the conclusion that a number 1 ranking as a best place to work was based on a singular, objective measure of what it takes to earn the vote. Not so.

An infographic created by Tommy McCall (McCall, 2011), who operates a data visualization studio in New York, revealed something really intriguing. Because of the dynamic way the data could be searched, the reader could dig deeper into a very important truth: the why behind number 1 and number 2 differentiated greatly. For example, the number 1 company on the 2011 list was SAS, a technology firm based in North Carolina. Employees valued the attribute of "care" and believed their employer provided "care" very well. Employee surveys said "SAS cares deeply about their employees."

But, the number 2 company, Boston Consulting Group (BCG), valued "people" the most. And the employees' reasons for giving their employer top scores was the value they placed on BCG's hiring of "smart, personable, and professional people."

Studying the list of winners and the attributes that made them winners, you could start to piece together a very interesting and significant story: different people are looking for different qualities in a "best place to work." The alignment between what employees value and the extent to which their employer delivers it is what sends a company to the top of the list.

Why does that matter? We're certainly not all striving to be at the top of a magazine's list as the acid test for how effective we are as leaders. However, we can learn a very valuable lesson: if we assume that highly motivated employees represent a strong part of our organization's ability to become a market leader, we will definitely want to instill in our own employees the sense they're working at a fantastic place. That being said, it makes sense for us all to be in touch with how our employees define a great place to work to see if we're on track.

The Container Store: Not Everything's for Free

You might think that pay and benefits loom large in what makes a job at a specific organization good. But in fact, people are far more likely to cite being a "member" of a "team." Which tells you a lot about the Container Store: one reason people enjoy their jobs so much is that they feel like they're part of a collective effort, working with people who share a common goal. The culture somehow really does instill a sense of shared purpose, which is perhaps the hardest thing for any organization to accomplish.

The words "team" and "client" often come up with the Container Store, which illustrates that the organization-focused firm also instills something of a client-focused culture—in other words, a sense of purpose beyond the individual job. But workers are far more likely to cite "people" and "benefit," which shows that the culture seems more about individual relationships with co-workers, and the cushy 401K matches make the long hours worth it.

Now, compare those corporate cultural norms with Google's. "People" again comes up often, but "team" isn't even on the radar; this isn't a hugely team-oriented culture. But the other words that loom largest are "free," "food," "perk," and "benefit." Google might be a fun place to work, and you might be among amazing people, but it's hard to escape the idea that the firm taps into a different satisfier for its employees—not the esprit de corps alone, but creature comforts like snacks and lattes being used as a way to build camaraderie.

What is the new profile for leadership that can serve at the helm, fulcrum, and hub, in today's organizations? What is the profile for the ideal leader?

The real imperative for business right now is to take the time to reflect on what type of leadership is called for in the new world order, and to make a beeline for any resources that can accelerate your path to embodying it. What can you do to come up to speed?

Find Your Next in Talent and Leadership

(🔍 find)

1. Ask great questions, ones that will keep you learning about rapidly changing competitive forces. Have you engineered feedback sensors into your organization to take the pulse of employee priorities? Look at other industries for ideas.
2. Embrace diversity. Spread leadership to new faces and new teams within the employee group. Incorporate a global view of talent, corporate culture, and leadership into your approach.
3. Rethink the human resources function to unleash the true potential of the people in your organizations. Revisit assumptions about the profile for the up-and-coming generation of leaders. Open up the channels of internal communication.
4. Master the rules for employee satisfaction in the new world. Does your organization need to reboot its culture to create a "best place to work?" Do you need to revisit your reward structure to align it with what drives today's employees?
5. Check for disconnects in your company's culture or remnants of past cultures that reflect old habits and former mindsets. Do you say high tech but act low tech? Do you preach innovation but reward consistency? Do you value diversity but reinforce uniformity? Does your brand seek to be responsive but you find that it operates within more rigid processes that don't allow you to adapt to the changing business landscape?

Chapter

7

Process Design

Empower Your Processes: A Systemic Approach to Company Growth

Your Process Design Self-Diagnostic

1. Processes can get boring and tired—and ineffective. How will you know when your processes need fresh eyes and a fresh look?
2. All processes aren't created equal. Which ones make a difference that your customers feel and appreciate?
3. Cleaner processes don't automatically mean a cleaner organization. How can new-and-improved processes push you toward the next wave of a competitive advantage?
4. Some of your processes might be fine the way they are, and others might be outpaced by the latest business crisis. How should your processes evolve to meet new competitive pressures?
5. Some companies seem to have figured it out. What exactly have they figured out and what can you learn from them?

Frederick Taylor may have been one of the first people to engineer processes, so perhaps that's why we can't seem to let him, or his process design, go. But we need to. His world—industrialization and workers—and ours—technology and virtual workers—are far too different. Businesses of industry in 1911, the year his classic treatise *The Principles of Scientific Management* came out, desperately needed that scientific approach to business operations. At the time, factories and mass manufacturing depended on smooth flow of material from point A to point B, making the ability to scientifically track efficiency with "time and motion studies" of paramount importance. And so Taylor helped companies in their shift from craftsmanship to rapid mass production. The approach taken by preindustrial tradesmen was no longer sufficient.

A lot has changed since then.

We're no longer in an era dominated by assembly lines, Taylorism, and efficiency standards (Taylor, 1911). Still, many of us are stuck in that efficiency-based mindset we no longer really need. Our process design, as a result, lags far behind our companies and our competitors. Assembly lines no longer cover the business landscape, so it doesn't make sense for the elimination of wasted steps to be the main measure of effectiveness. Assembly line-manufactured products put the output front and center, and producers who could maximize the throughput of their factories could reign as market leader, especially in industries like automobile manufacturing. But the conveyor belt gave way to new modes of operation that distinguish our companies from companies of the past. At some point between that era and now, new perspectives on process effectiveness have found their way on business agendas, all intended to hit at the heart of the central issues driving competitive advantage for every age.

Following Taylor in the transition from the 1940s to the 1970s, the world of business migrated from the factory to the office, and the definition of processes moved along with it, to include ideas like workflow design and paperwork simplification. By the time the '70s had upgraded employees from manual typewriters to word processors, office efficiency reigned king. Business writers saw office automation as the next frontier in process improvement, and rightly so, making the world ripe for Kinko's in 1970 (Graham, 1950).

But not every company was a good fit for equipment and work flow-centric systems of operation. Companies like CBS, number 17 on the Fortune 500 list in 1965 (Fortune Magazine, 1965), had to compete in categories like creative development and audience ratings in order to attract advertisers and needed to develop best practices that would integrate data like the Nielson ratings into their business processes. And market leaders like Procter & Gamble, number 24, needed something that would incorporate the view of the consumer into its product cycles.

By the 1980s and the dawn of the Information Age, a process design makeover was in order. The drivers of competitive advantage had changed. W. Edwards Deming pioneered the shift, applying the lessons he learned in U.S. war plants and Japanese manufacturing firms to make a case for quality control, a strategy he described in his 1982 book, *Out of the Crisis* (1982). That gave way to Total Quality Management, or TQM, a process meant to hold everyone—not just the executives—responsible for a company's competitiveness.

It was no longer just about industry. The Baldrige awards—presented to companies that exemplified "best in class" processes and performance—were established in the late 1980s, when process design had been heralded as a tool for competitive advantage for not only manufacturers, but also for the growing ranks of service-oriented companies like Xerox Business Services (winner of a 1997 Baldrige Award) and Sharp Healthcare (winner of a 2007 Baldrige Award). On or off the assembly line, these systems and modes of practice mattered. TQM and Baldrige helped cement the reality that process was a key factor leading to a competitive advantage, and that workflow, people, and service now functioned as *components* that could drive a company's competitive advantage. Process wasn't just a way of getting things done anymore, but a result.

Then Motorola devised the Six Sigma system, a discipline designed to reduce manufacturing defects, taking us back a step or two. It was well-received, even inspiring leaders like Jack Welch at General Electric to forcibly drive his company's own manufacturing processes and service delivery to new levels of effectiveness. And though the process is used in a number of ways today, Six Sigma, like the ones before it, still found its inception and relevance in industry and manufacturing.

That began to change for good.

In *Reengineering the Corporation*, Michael Hammer and James Champy declared the business world ripe for a "revolution." And they were right. Their "manifesto" set out to take on urgent competitive challenges marked by a newly computerized and interlinked world that was characterized by a wave of new entrants—dot-coms and global players and every genre of company in between—to the world of commerce (1993).

Companies then and now find themselves in the midst of a dizzying sweep of competitive challenges that demand better processes, ones that can grab the attention of every success-driven business leader. Welcome to the Information Age, when processes travel at computer speed, and process design itself can give usa competitive edge. Now we have to embrace a new attitude toward processes: rather than focusing solely on efficiency or best practices, our processes must be ones that allow us to adapt to big shifts in the competitive landscape. And they have to enable us to sense what lies ahead.

It's not enough to design processes that are efficient. They also need to equip us with faster reflexes. Ours is a business environment in which big shifts happen without much warning. As a result, company leader shave to keep their eyes on the big picture to stay afloat. Our process improvement can't simply rearrange the deck chairs on a sinking ship. It must avoid the consequences of the icebergs up ahead. Think of the number of telecommunications companies taken down by deregulation, then the ones that survived and triumphed by redesigning processes with the new realities of cable television and wireless connectivity in mind (Shaw, 1988). Remember that ours is a world in which one minute a company like AT&T can bemoan the infrastructure challenges of the "last mile" of installation, busying itself with streamlining its growing backlog of orders, and in the next its landscape can change, opening its industry to new players and innovative technologies that make those time-saving processes null and void. Before anyone could see it coming, wireless data moved telecom companies past the place where land lines were still relevant. Consequently, the companies that didn't shuffle fast enough to keep up were left in the dust.

IBM makes a good example of this. Think of the families that spent weeks planning an extensive wiring system to connect their entire house to a central computer server from every room, only to wake up one day to realize a wireless router solution would have saved time and money over a hard-wired blueprint. They might have already invested in a very detailed process to prioritize the wiring configuration, set budgets for a range of scenarios, interviewed vendors,and developed project flow charts for implementation. Right on the cusp of pulling the trigger, wireless connectivity could come on the scene, completely upending that list of options. Once the poor team from IBM's Smart Home wired division wakes up with that same realization, they'll understand that their expertise and product line will soon hit the dusty shelves of obsolescence. If IBM were to judge its success based on the process efficiency of its initial design, they'd get an A, but if they looked at the broader context of lasting relevance, it wouldget a much lower grade. Every company faces the threat of being caught unaware of the solutions that will revolutionize its competitive game. We all risk ending up like Indiana Jones' swordfight challenger in *Raiders of the Lost Ark*, using our fanciest sword skills during a fight, only to be completely disarmed by our challenger's novel solution—a gun. Processes may not be able to help guard against that kind of shape-shifting innovation, but they can get you ready to face the competition, what's facing you now, if you've got the right stuff.

The New Realities of Process Design

1. Complexity can mean opportunity. Armed with the right redesigned process, a company can use it to get ahead.
2. With so many processes to focus on, we need to make sure we're focusing on the right ones. Not every process, even optimized, will lead us directly to growth.
3. Process redesign can do more than simply enable good products to get to market more smoothly. Processes can create competitive advantage.
4. The process can *be* the product.
5. The signs of our future already exist in adjacent and even unrelated industries, and they can provide insights into consumer preferences.

1. Complexity can mean opportunity

Envision the role of an interior design firm a decade ago. A well-trained individual with a flair for house and home could blend a client's objectives with materials into a finished interior. The only problem is a matter of scale, and a matter of process. Even a top interior designer would be limited by those processes that required his or her time and talent to create the vision and drive the workflow. Process redesign, supported by recent technology advances, solves that problem. Products like Decorator in a Box and Design in a Box provide scale and market reach to interior designers like Annie Pauza and Nicole Sassaman. The

Design in a Box: We Come to You

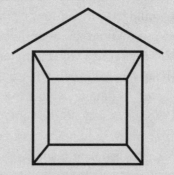

More efficient processes translate to companies finding newer and cleverer ways to meet their customer demands, no matter where they are. LRM Interior Design did just that with a new concept that is popping up in design service providers all over: Design in a Box. The company simply gets the prospective customer to fill out a form about what he or she wants and needs in terms of interior design, and keeps adding boxes as long as the client has rooms to pretty up. And, as an added bonus, customers are more active in the process, no longer just handing over the keys and floor plans without a second thought. The deliverable is a map and the client follows it. Because everything is by recipe, the concept works near and far, and it can be done remotely with photos and some captions, and a phone call.

"in-a-box" models shift the design process from being largely face-to-face, live interaction, to a relationship enabled by online resources. With Design in a Box, a client gets a choice of a predesigned or custom-made room. Some design boxes can be shipped right away, or if customized, they can be shipped in four to six weeks. The box includes everything a client would need to put the design into place in his or herhome: conceptualization, a material list, a furniture plan, and step-by-step instructions.

Large companies can also drive new value to their customers with the same kind of push toward scalability. In 1999, IBM's Chairman's Award, which recognizes customer value, went to its procurement process team for its tremendous success in reshaping procurement into an e-business, a complex, aggressive undertaking that was critical to IBM's ability to compete in a faster-paced, Internet-based world of commerce.

According to CEO Sam Palmisano, "Purchase-order processing time went from one month to one day. The time it takes to get a contract in place dropped from six to twelve months. Average contract length shrank from forty pages to six. The team completely transformed the procurement process, which had been paper-oriented and very local with no economies of scale or ability to negotiate. As a result . . . our rate of maverick buying outside the process, leading to bad terms, dropped down to less than 2 percent from 30 percent. Internal customer satisfaction went from 40 percent to 85 percent" (Hammer and Champy, 1993, p. 196).

Eleven years later, IBM leveraged its insights in process improvement into a core corporate offering that centered on streamlined processes as a driver of market differentiation. It set out on an acquisition and expansion course to bring those services to market, purchasing Sterling Commerce, a cross-channel commerce software provider. According to a general manager of IBM, "The combination of IBM and Sterling Commerce enables the integration of key business processes across channels and among trading partners—from marketing and selling to order management and fulfillment." Not only did IBM reengineer its processes, it invested in them (IBM, 2010).

Today's market dynamics offer virtually every business, whether midsized and entrepreneurial or larger and more complex, a channel for streamlining processes that strengthens the important components of the customer relationship and automates those aspects of the process that customers view as a commodity transaction.

Every step of the customer experience from marketing and sales, to order fulfillment and customer service, offers opportunities to optimize processes while increasing customer impact. Companies like CarMax and WalkScore have streamlined comparison shopping on the Internet, while organizations like Siemens have changed the nature of shipment tracking with RFID, allowing customers to keep an eye on their purchases at every point in the process, from manufacturing to delivery.

At the same time, the internal processes that support the customer experience, including logistics, purchasing, management decision making, workflow design, and manufacturing, have undergone huge transformations akin to IBM's award-winning purchasing team's process redesign. With a strong track record for streamlining logistics, large companies like DHL have also stepped up their process improvements to a new level of innovation, through a dedicated solutions and innovations practice that builds process improvements into the heart of their competitive strategy. With creative "focus projects" like Cold Chain and City Logistics, DHL looks ahead in fields with temperature-sensitive shipping (like the life sciences), addressing quality of life issues related to shipping in large metropolitan areas.

A company's processes are no longer the unloved step-child of edgy business strategy; they've moved to the front burner in impact. Process has the power to propel a company from near death, in which it gasps for air just to keep up with market demands, to super-athleticism, a sought after position of mastery over the competition.

2. Not every process, even optimized, will lead directly to growth

Open six office doors and eavesdrop on six conversations about a process and you will no doubt experience a modern version of an old story—all six will be talking about the same company, but you won't be able to tell it. Like the story of the blind men and the elephant, executives tend to look at a process from their own, sequestered points of view, and each perspective is too incongruent for them to make much sense. One reason for the misalignment is that business processes can't be one size fits all: processes are generally designed with a single point of view in mind. For example, an airport architect could draw up a blueprint with a list of several goals in mind—a user-friendly passenger pickup area, a shorter length of time for baggage retrieval, or a shorter distance between gates—but a single design won't solve all of them.

Sometimes, as was the case with the Denver International Airport, a well-intentioned, strategic reengineering of baggage handling doesn't deliver a more user-friendly solution (Weiss, 2005). There, the airport's design team invested time and talent in a high-tech system for streamlining baggage handling that featured bar code tagging, remote controlled carts, and a 21-mile-long track, all intended to enhance tracking and delivery, only to discover that the glitches in the newfangled process caused more mix-ups than efficiencies.

Airports are the greatest challenge in processes. The Charleston International Airport went through a bevy of changes, mostly in an effort to allow for larger concourses and baggage claim areas. Based on preliminary reports, the airport personnel saw a great savings to passengers, and greater income to the airlines, with room for an extra 200,000 flyers. That, of course, also means more places for all those people to go and sit, and even park. It would even make the airport hub-friendly.

Airports also have to deal with the unexpected—like increased security requirements post September 11—when old design assumptions take the back seat. Reengineering a process designed around one strategy, like a short walk between airport entry and departure gates, can collide with others, like keeping people safe. Retrofitting an airport to achieve the new objective of security can prove almost impossible. Existing airports struggle with bolt-on fixes, while new airports have the advantage of being able to integrate security requirements into the initial design.

To their great benefit, the designers of San Francisco International Airport's new terminal evaluated the new security requirements before they even broke ground. Starting with a blank slate, and armed with their analysis of customer frustration regarding the new Transportation Security Administration procedures, they came up with innovative, traveler-friendly elements, including a "recomposure zone" that provides a peaceful area for putting on shoes, repacking mobile phones and laptops, and getting organized before heading toward the gates (Arieff, 2010).

You can imagine the people around the planning table every time an airport's processes are on the agenda: financial analysts, customer care specialists, marketing and public relations personnel, engineers, vendors, architects, customers, baggage handlers, airport retail representatives, pilots, and virtually every other party affected by the redesign. Each person could easily argue for a solution that would perfectly address his or her individual set of priorities. How can companies adopt the ideal perspective on business process when there are so many to choose from? Which is the one to unlock a competitive advantage?

Airport Design Illustrates Several Universal Truths Underlying the Most Strategic Approach to Business Process:

1. Processes can't be optimized around every point of view.
2. Process design around one factor (traveler comfort) can interfere with another (security requirements).
3. Redesigning a process might begin with add-on fixes that only temporarily address the symptom at hand (added security lines post September 11).
4. Add-on fixes can resemble a game of Whack-a-Mole, in which a company never gets to the root of the underlying problem.
5. Redesigning processes from scratch is expensive, but sometimes it is the only way to truly achieve your strategic goal.

3. Processes can create a competitive advantage

When we're looking for ideas to drive market leadership for our companies, we sometimes place process design behind discussions about more tangible assets, like the products we produce. We shouldn't. We need to recognize how some companies—Netflix (movies delivered by mail or online), Avis (rapid return: self-service check-in), and Siemens (radio frequency identification chip tracking)—have become market leaders of process, by introducing a new industry standard for a way of doing business. These "process leaders" have proven that process improvement can actually drive competitive advantage.

Johnson & Johnson: Looking to Logistics for a Global Edge

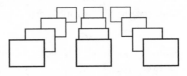

Herwig Maes, director of strategic sourcing and supplier relationship management for Johnson & Johnson, reveals his fascinating journey to rework logistics on a global basis. His examples illustrate how he connected with peers like Baxter and GlaxoSmithKline to take a strategic look at shipping, packaging, inventory, and virtually every process in getting health and medical supplies from the point of manufacturing to the point of distribution—for consumer goods like baby powder or mouthwash—or the point of care—for medical devices, pharmaceuticals, and diagnostics. Once Maes sorted through how various goods were handled, he made some big revelations. First, he realized there were opportunities for enhancing customer satisfaction by re-categorizing and re-grouping some products by customer need: for example, refrigerated items frequently require special handling, and time-sensitive items like specialized surgical supplies could be grouped separately for faster delivery.

Herwig discovered that the standard lessons of Six Sigma required a new layer of insight. Certain considerations—like how a box of a certain size might contain baby shampoo, or vials of vaccine—challenged the traditional approach to process optimization. Maes and his colleagues knew they had to reevaluate their company's enterprise-wide supply chain and apply the lessons they learned in Europe on a global basis, beginning with fundamental questions like, "How do the internal logistics of our hospital clients work and what can Johnson & Johnson do to deliver exactly what the customers need when and how they want it delivered?"

They moved fast on the learning curve by studying global companies in other industries like Nike, DHL, Skil (a power tools company), and Black & Decker. According to Maes, "It was a bit unusual to study Nike since our products are so different, but it turns out that Nike has a huge European distribution center in Belgium that incorporates barges and taps into the sea freight waterways. We looked at DHL's capabilities in the areas of RFID (radio-frequency identification) and temperature control, and realized we could add tremendous value to our customers by enhancing our tracking and data transmission and also building in innovations related to our cold shipments. For example, we integrated temperature readouts when those shipments are loaded onto the truck, aircraft, or van, or placed in the warehouse. We also had a great 'a-ha!' moment when we learned from Black & Decker and Skil about warehousing of spare parts and just-in-time delivery of a specific power tool to a construction site. If that approach works for drills, why couldn't it work for our global distribution of surgical supplies?"

Many strategy sessions focus aggressively on process as the element that could either make or break a company's ability to gain market leadership, such as a large telecommunications company and the best redesign of a customer call center's work flow, ora sandwich shop chain and the queuing of its customer lines for shorter wait times. In each example, the teams decided that their competitive advantage hinged on process design and implementation. They learned that the best way to design a game-changing process is to define a single strategic driving force for growth and align every process to support and enhance that one variable.

In one specific example, a securities firm I worked with was tasked with increasing customer retention, at a time when the world of selling and managing securities had evolved into a very level competitive playing field. A Six Sigma–based analysis revealed a long list of process issues, but it would have been too expensive to address them all. Instead of going after every single one, the company uncovered the processes that would have the biggest positive effect on what mattered to them right then—customer loyalty—and tackled what they felt was at the heart of their customers' satisfaction, simplicity of statements. Customers were happier, and the securities firm grew.

The firm gained an important insight for times like today's, when nimbleness and customer responsiveness can drive competitive advantage: "Measure twice, cut once." When process is intended to strengthen your competitive edge, work on why any given process improvement is on the drawing board in the first place (in this case, better customer retention), then tailor the fix to the outcome.

4. The process can be the product

Sometimes an opportunity for a new business emerges by piecing together components of one industry and gluing them with shards and elements from another, thereby creating an entirely new category. Dallas clothier J. Hilburn is an example of exactly that type of business, one that serves as a model to any company looking to rethink the processes that ready a product for market and the systems that nourish loyal customers.

J. Hilburn: A Retailer with a Little Dell and a Little Avon

J. Hilburn uses processes to create a competitive edge and it is nothing short of cross-industry. The innovative clothier sought to make custom-made clothing more workable, using the direct sales model of Avon and the customability options of Dell. Of course, it's simple to order from them online (like with Amazon) and the company's process management rivals Toyota. Its sales staff go to the customer's place of work or home, the shirts are ordered from China and they are a lot less expensive. Style advisers make commission on whatever they sell after some initial expenses, and they receive a percentage of the sales by any new advisers (with a maximum of five) that they bring to the company. They can't avoid the process inefficiencies that come with anything custom-made (i.e., the way the clothes are cut and sewn), a completely different, individuated process from that of store-bought clothes. That means the company makes far less clothing items in a day—only six. But still, for people who would like a tailor-made clothing option that won't break the bank, J. Hilburn is the place for them.

In the case of J. Hilburn, the team had the advantage of being a start-up, with no road-worn habits and few preconceived assumptions about how shirt buying should be done. But its fresh take offers valuable lessons on how to achieve better yields from existing processes and approaches.

But Wait: Paperless or Electronic?

Under some conditions, making changes to processes can be more challenging than making changes to products. Processes that have dominated the way we do things come with habits that can be hard to break. Sometimes those habits are so deeply ingrained we don't even realize we're stuck.

We need to be on the lookout for those old patterns to make sure our new processes break them for good.

A large chemical manufacturing company I once worked with was finally ready to go paperless. Even its most seasoned plant manager, a guy who knew all of the ins and outs of running the facility with a trusty set of redlined blueprints he was never seen without, appeared to be on board. As the company moved forward in its "paperless office" initiative, he couldn't have been more supportive and enthusiastic, leading his team through the transition with tremendous charisma.

The paperless system went live, and the usual round of celebrations marked a successful completion of the multi-month project. The plant manager seemed ready. But he wasn't. A week later, he could be found leaning over his drafting table, and not his computer, to analyze one of the components in the system that had gone down the previous evening. He hadn't made the shift.

Process redesign, while powerful, requires adjustment in order for its users to truly cross the Rubicon and fully embrace a new way of doing business.

Find Your Next in Process Design   (Q find)

1. Assess what's missing from your company's performance. What signs are telling you that your processes might be holding you back? Ensure your improved processes can measure success. Whose perspective should lead the shift to a new process? Why?
2. Take in the current competitive landscape. How have standards changed: speed to market, online experience, warehousing, logistics? Where will you get the most payback for an investment in process change?
3. Scan the future before starting.
4. Align your team with the most strategic opportunities for redesigning processes. Creating efficiency for efficiency's sake is inefficient. Check to make sure that the change takes hold. Remember old habits die hard.
5. Look to other industries for processes that are driving their competitive edge. Learn from the exampleof J. Hilburn and borrow processes from the best in practice.

Chapter

8

Secret
Sauce

Keep Them Guessing: Finding the Right Formula for Your Secret Sauce

Your Secret Sauce Self-Diagnostic

1. What is it about you that keeps your customers coming back? Do they see all that you are and all that you offer?
2. Who stands a chance of stealing those devotees or encroaching on your competitive space?
3. How can you stay wired to the market pulse, even as it vacillates?
4. What will make your company a market leader in the next year or two?
5. What's the secret to differentiating your brand in today's competitive landscape? Have others in industries outside of yours figured it out?

The sad fates of the soon-to-be-extinct threaten us all. Those endangered and once powerful species comprise a long list: the U.S. Postal Service, Yellow Pages, classified ads, video rental stores, dial-up Internet access providers, VCRs, ham radio, answering machines, incandescent bulbs, stand-alone bowling alleys, milk delivery men, hand-written letters, honey bees, wild horses, personal checks, drive-in theaters, mumps and measles, printed magazines, TV news, analog TV, the family farm, ash trees, and even Chesapeake Bay blue crabs (*Twenty Five Things About To Go Extinct in North America*, 2008).

Here today and gone tomorrow are entities that were one-time leaders in their fields. Now they are facing extinction after a great run as companies, brands, services, and even species (like the honeybees). At one time, they rode the wave of popularity and thrived without threat, and now they find themselves in a downward spiral with little hope for recovery. They didn't think it could happen to them, just as you don't think it can happen to you. But don't mislead yourself. Instead, protect yourself. How can we safeguard our companies from joining the others on the list?

Any time we think we've cemented our popularity or established bulletproof brand distinction, we could still be at risk for losing ground. The signal can come suddenly or leisurely, and it's up to us to respond. It's up to us to be ready. It might be the emergence of a new, better-equipped, far nimbler competitor that grabs market share; new products and technologies that make ours as obsolete as pay phones and land lines; or an unforeseen circumstance or event that poses a threat to how our company is perceived in the market place like a tampering scandal or an oil spill. It could be anything.

Maintaining a point of difference—or secret sauce—is not out of your control. But we also don't have the luxury of waiting for tomorrow's newspaper to tell us what we need to do—or what we should have already done. If we have a hunch that the market is about to shift or early warning signs are telling us that our secret sauce is about to lose some umami, we must act quickly.

We have to achieve a balance between struggling to stay ahead and lounging in benign neglect of the competitive radar screen. How often should we look over our shoulders for the other runners in our race? How far on the horizon are the new forces that could blindside us?

Wake Up Calls Come in Many Forms

We have a lot to monitor. Small hiccups in performance sometimes don't add up quickly enough to equal the risk we're facing—eroded market share, low customer satisfaction scores, dried-up margins. Raised red flags, like highly public crises, can shock us more dramatically into high alert. No one is immune. Even long-standing brands and companies like Tylenol or BP are fragile when it comes to public perception. Those crises can have hidden benefits as in-your-face warning signs that your company needs to step back and rally your employees to be more attuned to the public. That sense of alarm— when everyone knows the stakes are high—can either lay a company's brand to rest (AirTran), give a company a chance to recover (BP's 2010 earnings recovery) (BP, 2010), or stimulate a corporate rebound (Tylenol's 1982 tampering crisis) ("Tylenol's Miracle Comeback," 1983).

Why wait for an emergency to try to make repairs? Why not send out sensors early to test the market? Why not harness the energy of market feedback for what could become points of distinction? Why not find your own secret sauce?

Secret sauce is the combination of intangible qualities like reputation and market perception that allows a company to maintain market leadership, become the gold standard, and lead the pack.

Sometimes we think we're selling the beef, but our customers are buying the sizzle. When Charles Revlon said, "In the factories we sell cosmetics; in the stores we sell hope," he crystallized Revlon's insights about the fundamental distinction between two types of makeup providers: one that sells a bottle full of chemicals and one that relates with customers and improves their lives. The drivers of that distinction might be hard to pin down, but they explain the difference between a combination of ingredients—sugar or high fructose corn syrup, caramel color, caffeine, phosphoric acid, coca extract, kola nut extract, lime extract, vanilla, and glycerin—and the global brand of Coca Cola.

Secret sauce is defined not simply by the ingredients that make up our brands, but also by the credit our customers give us for the whole recipe. What tips the scale in favor of our brands?

Brands matter, as well as what keeps them popular. Those factors are fair assessments of how our companies are doing at keeping up with market preferences, and represent more than a whimsical thumbs-up or thumbs-down from the public. The first place we need to look to bolster our brands, especially when we're not in crisis mode, is what they represent. We need to take an objective look at whether the core that supplies our secret sauce still works.

A pioneer I worked with—a leader in the field of industrial psychology—had invented an innovative and well-researched personality assessment, similar to the Myers-Briggs diagnostic. The core of the company's differentiation and brand leadership during the 1980s and 1990s was the standardization of its reports, sort of an SAT scoring system for employees. During those early years, the format for the reports held tremendous appeal for human resource directors and other executives tasked with hiring new talent that would perform exceptionally well. At that time, scientific certainty was attractive when human resources groups were looking for uniformity; a highly standardized report made a lot of sense. Sales of the assessment were steady.

Times changed and, beginning in the late 1990s, there were signs that the cultures of the Fortune 500 clients were starting to shift, threatening the assessment company's central place in the recruiting and promotion of employees. The credit that customers had once given the diagnostic assessment company for its objectivity and report consistency was eroding; the secret sauce of dependability wasn't worth as much as the new customer preference of customization and flexibility.

Early cues were hard to read and appeared to be merely ebbs in normal sales patterns. But gradually the company realized softer sales were a sign of something more significant: comprehensive reports were no longer the tool of choice. It looked on as its clients moved from a mainframe technology to a more flexible and customizable PC standard, and demanded flexibility from its vendors that followed suit.

By the late 1990s, the company's customers rejected what had once been their prime point of difference—the gold standard diagnostic—and corporate leaders demanded a plug and play version of the reports so that individual supervisors could slice and dice the provided insights to fit their unique recruiting and assessment requirements.

The brand was forced to transform its main product from an SAT-like score to an "Intel-inside" version that integrated the reporting results into a wide range of customized formats. Over the course of a few months, the company reached out to itscustomers and dove deep into how the employee insights were being applied, discovering that the original intent of standardized scoring was overdue for a tune-up, one that would ride up the corporate on-ramps in many areas: hiring, performance review, team engagement, talent development, succession planning, outplacement, and even competitive benchmarking.

Luckily the company caught up quickly and shifted gears to meet the new requirement for flexibility, boasting an extensive line of customized reports, interactive apps, and Web-based products that matched the new customer requirements. But not all brands have fared so well. Consider the latest in the rivalry between Coke and Pepsi, and Pepsi's innovative and promising decision to swap Super Bowl ads for social media coupled with a viral campaign related to social responsibility. But, when the "jury" weighed in by way of sales figures, no credit was given to the brand tweaks. Coke still came out ahead.

2010: Diet Coke becomes the second most popular soda, taking Pepsi's long-held spot behind Coca-Cola's Coke. Coke holds strong at number 1, with 17 percent market share and 1.6 billion cases sold. Diet Coke comes in second place, with 9.9 percent market share. Pepsi has 9.5 percent market share, an almost 5 percent decline from 2009.

It's the Goldilocks dilemma in action. We have to figure out which approach is "just right" and wake up in time to save our brands from a possible downward spiral. At the same time, we have to make certain that our fixes earn credit with our customers, no small task.

Peter J. Wright, a seasoned executive with more than two decades of experience of global brand development (with brands like Unilever and Estee Lauder, AIG, and Zurich Financial), has an insightful theory on why successful companies cling to old mindsets, even in the face of factors that dramatically challenge the long-term staying power of the status quo. He tells a great parable of three young women in a car driving downhill on a dangerous, winding, mountain road (Wright, 2011).

The girls see the stereotypical bad boy coming up the hill toward them, rounding a hairpin curve as he approaches. The boy is almost a textbook specimen of obnoxiousness, complete with a red convertible sports car, designer shades, and blaring music. As he comes up the hill full speed toward their car, all the women come to the same conclusion—he's a jerk—and scream hysterically as their driver swerves to avoid a potential collision. In that split second, the jerk yells back, 'Pigs!,' which to them is the last straw. They conclude they were right. He's a jerk.

And the women's assumption at that instant prevents them from fully grasping the intent of the jerk's outburst.

As they round his same curve, still gasping for breath from the near-hit, an actual herd of pigs appears (and the women are forced to reassess their snap judgment of moments before). What they saw was a convertible, what they heard was loud music, but it didn't mean the man was a jerk.

Peter's story applies to corporations. Too many companies dismiss new market conditions as temporary or attribute declining sales to a poor economy or think that new competitors will pose no threat. Their leaders are latching onto what they want to believe the facts are telling them by clinging to their assumptions of where the company is headed, regardless of evidence that screams of a more meaningful interpretation. We've all lived through our own moments of certainty about the relative non-threat of a newcomer to a competitive set, only to be caught off-guard as the competitor enters that arena and makes an impact. We allow ourselves to be lured into complacency. We hold tight to what we believe to be true (or, in the case of the jerk in the sports car, our existing paradigms). And we miss the future that is right in front of us.

But some companies do wake up to the realization that times have changed and tap into the wave of that change.

Stir, Stir, Stir: The Recipe of Success for Today's Hotels

Daniel Edward Craig, hotel consultant, speculates on what's next for hotels and guests (2010):

1. Lowered price integrity. Hotels will keep dropping rates, without taking away any of the creature comforts we've grown to expect. Free breakfast barely needs to be mentioned.

2. The experience of the new bargain shopper. Customers will want lower rates and better service.

3. Packaged pricing. To increase prices, hotels will add more and more services to every room, in an attempt to hide the price hike.

4. More spending on the inconsequential. Luxury goods and services, including expensive hotels and unbelievable amenities, will be in vogue once again. But only for a select few.

5. A hotel for your lifestyle. More boutique hotels will open that offer a "greener" experience, while staying cost-effective.

6. Hotels on Facebook and Twitter. Staff can talk about guests, too, while guests talk about the staff. And all in an open forum.

7. Being everywhere at once. Guests will do everything at home that they once had to do at a hotel, and so, they might not need such nice accommodations or any at all.

8. A focus on safety and health. Hotels will begin to check for disease and terrorism at check-in.

9. Go to the mattresses. The bed still matters and they'll keep getting better, and softer, and thicker, and fuller.

10. Meetings will be more about work, and less about location. In order to save, those big annual meetings won't be anywhere very interesting, so lesser-known locales will have to step up.

Hotels are finding they need to try harder than they ever have before. A cushy bed and a high-ceilinged lobby won't be enough.

The difference between becoming irrelevant (and risking extinction) and thriving (and adapting) can boil down to a brand's ability to snap to attention, greet change, and reinvent itself to meet the needs of the new elements in our lines of sight. We need to see the pigs and not the guy, then adapt accordingly. In building brand distinction, we need to take into account the new realities of corporate identity.

The New Realities of Formulating a Potent Secret Sauce

1. Yesterday's success doesn't necessarily hold the secret for sustainable brand leadership.
2. In today's cluttered competitive world, creating a new killer brand or re-energizing a fading brand requires a jolt of innovation and energy.
3. Brand distinction in the Web 2.0 world includes new factors: bottom-up brand invention combined with breakneck speed.
4. Brand space is crowded; seek out new territory.
5. Spin the globe: the reality is that every brand is touched by global dynamics.

---

1. Yesterday's success doesn't necessarily hold the secret for sustainable brand leadership

Jeffrey Immelt, G.E.'s chairman and chief executive since 2001, saw his "pig" for what it was—a brand that had lost its secret sauce and was on its way to losing even more. Long-standing success coupled with a stellar run at profitability couldn't guarantee future leadership for the G.E. brand. Nothing could. When Jeffrey Immelt took over the helm in 2001, G.E. was at the end of a great run of success and innovation. The care and feeding of G.E. Capital had taken the company on a new adventure, with a focus on ratios, returns, and financial measures (not production, R&D, and invention), but that good fortune had defocused the company away from its roots of manufacturing.

By 2005, Immelt began the pursuit of a brand differentiator that he believed would sustain the company in the longer term, writing a new story for a brand that ran counter to the notion that innovation has to be new and different. Immelt determined that the sexy secret sauce for G.E. would come from a revamping of its roots by doing a whiz-bang job of rethinking manufacturing, with a bent toward energy efficiency. He called the initiative "ecomagination" (see more about ecomagination on page 172).

"The G.E. campaign to promote energy-efficient products, begun in 2005 and called ecomagination, is a model of Mr. Immelt's efforts to push large-scale change. It began because he believed that energy efficiency and alternative energy sources were growth markets and that G.E. should capitalize on those trends" (Lohr, 2010).

While that decision may still be in its early stages (without long-term revenue data to tell how well it worked), the path that Immelt took serves as inspiration to those who find themselves at the end of a revenue road, and in the timely place to bolster their brands. He looked at the long view on offerings that would resonate in a changing world—where energy efficiency would reign—and bet the future on the company's ability to lead the pack in delivering novel products and redesigned manufacturing processes.

Brands can go stale one notch at a time, but if brandmakers are lucky, they can catch themselves in time to rebound. And, not every company has to be a giant like G.E. to invest in a reinvention that can spice up its secret sauce. Whether a local car repair shop offers customized car care communications, online appointment scheduling, and text message status updates, or a community hospital features iPad-based concierges to navigate its services, the ability to re-establish our brand's points of difference are within reach. It starts with rethinking relevance.

2. In today's cluttered competitive world, creating a new killer brand—or re-energizing a fading brand—requires a jolt of innovation and energy

Today's market opens the door for the creation of powerful cult brands and tribes of supporters. Consider brands like Toms Shoes. It was founded by Blake Mycoskie, one-time runner up in *The Amazing Race* and 34-year-old serial entrepreneur, a man who translated his passion for social causes into a revolutionary for-profit shoe company that has earned more than $13 million in sales within five years. Mycoskie commands standing room only at conferences like South by Southwest (SXSW) that attract innovative entrepreneurs, and he has built a stellar team with highly credentialed people who have worked for free (like Candice Wolfswinkel, the chief giving officer). He has partnered with Ralph Lauren and the Clinton Global Initiative Conference. He has taken his brand to true cult status by giving a pair of shoes to a poor child every time a pair of shoes is sold.

Not every brand has it in its DNA to become a cult, nor should it. But in days when brands are less about what they tell you they're about and more about what the company stands for, companies like Toms Shoes connect with customers in a deeper way than most other companies even try to.

Matthew Ragas and Bolivar Bueno call it "cultrepreneurism" in their book, *The Power of Cult Branding*, and cite Oprah's Book Club, Linux, and Harley Davidson as examples of brands that have built followings above and beyond their competitors in driving customer passion (2002). Seth Godin describes the phenomenon in which people want to belong

Seven Golden Rules of Cult Branding

1. Cult brands satisfy the consumer want to be part of a group that's different.
2. Cult brand inventors show daring and determination.
3. Cult brands sell lifestyles.
4. Cult brands create evangelists.
5. Cult brands always create customer communities.
6. Cult brands are inclusive.
7. Cult brands promote personal freedom and draw power from their enemies.

to a group, ascribe to a set of values, or follow a larger purpose that connects them to other people like themselves as "tribal." Godin emphasizes the importance of a leader who emerges to galvanize the forces from that tribe and translates it into a brand with "legs," and the staying power to attract a loyal following that responds to the secret sauce stirred in by the brand's founder (i.e., Howard Shultz from Starbucks). According to the tribe theory, "Leaders of those tribes emerge, most often from unlikely spots—not the conventional paths. The leaders see something different and set forth on a path to bring others to their perspective" (Godin, 2008).

If your company is poised to shift the thinking of a large group of people with a brave new approach or a new story (like Toms Shoes), or it is prepared to rethink a cultural landscape (like Starbucks' "third place"), you could be in a position to transcend the traditional brand mindset and enter cult leadership territory.

3. Brand distinction in the Web 2.0 world includes new factors: bottom-up brand invention combined with breakneck speed

John Winsor is the founder of the crowd sourced agency Victors & Spoils and co-author of *Flipped, Baked In*, and *Spark*, three edgy books about the new world of the brand, a world where co-creation, customer engagement, and bottom-up thinking move to a company's front burner as a way to create secret sauce, reinvigorate a brand, and spread the word like wildfire.

Winsor's fundamental belief is that today's market is driven by an entirely new market force, something only possible by the speed, ubiquity, and dynamic of Web 2.0 connections and the transformation of availability of technology. In a recent conversation, Winsor talked about then and now: how before a precious few ideas were held by a small cadre of "creative" types and now video cameras, blogging tools, music production capabilities, and social media sites are open to everyone.

According to Winsor, "when 'Charlie Bit My Finger'" gets as many hits as a high-end Lady Gaga video, you know you're experiencing a new dynamic in creativity" (Winsor, 2011). Companies will fall behind if they can't harness the power of these new forces.

The upside of the Web 2.0 world is that there is a lot more creativity to tap into. The downside is that our companies have to interpret signals from many more sources: Facebook groups, purchase patterns, geofencing data, mobile device trends, bloggers, new media, and niche markets that can speak for themselves. Talking to "many" as a generic target customer base won't cut it. The small groups of people who relate to our brands have individual tastes, voices, and forums to discuss our companies and products, generating content about us whether we insert ourselves into their conversations or not.

The book *Wikibrands* synthesizes many of the discrete, independently developed tools and trends that evolved during the first waves of the social media revolution. The authors outline a wide range of market insights that are available in a "wiki-world," in which a huge network of individuals connected via the Internet can provide extremely useful insights into our brand's differentiation. They share a host of practical examples set by companies that have taken advantage of the new shift toward information transparency. We can all learn from the Harley-Davidson HOG community of fans and the company's fanning of its flames of passion, as well as Dell Computer's migration from "Dell Hell" in 2005, when its customer ratings tanked and it later emerged as the second most engaged brand, largely attributed to its highly focused social media strategy (Dover, 2011).

Today's Internet-based relationships provide a new mechanism for aligning our brands to shifts in the market, but the stakes are higher for every business. Transparency makes our messaging more than simply a one-way push of information related to our brands; the new speed means that if we blink, we might miss a crisis and even a competitor's newfound intimacy with our customers. Whole industries, like the world of publishing, have been migrating away from an almost-sacred status with its customers to a much more fragile one, or worse, perhaps even extinction (James, 2011).

4. Brand space is crowded; stake out new territory

There's a theory that if we have five restaurants on our favorite list and we add one to the mix, we'll let go of one of the original five to make room. We won't add a sixth. Ours is a finite brand space. Scott Wilson has a similar theory about the wrist that provides a tangible visual for the concept of brand space. Scott asks the question: "Who will 'own the wrist' in the coming decade when Gen Y-ers have traded their watches for iPhones?" Or perhaps that time has already come.

Scott thinks the wrist is wide open territory for the grabbing. If his recent success story of pre-selling close to a million dollars worth of watches in one month on Kickstarter is any indication, his theory about the wrist might open a door for Apple, AT&T, Netflix, or any company to come in and create a reason for us to put something more than a simple watch on our wrists.

Many competitive landscapes are rich with opportunities for a new carve-out. Starbucks identified an untapped desire for people to hang out somewhere that wasn't either their home or their office, but a "third place." Netflix (and others who have followed in its path) did it with streaming videos. And one of the latest sensations in the business world, Groupon, managed it with a fresh redesign of the coupon clipping options from last generation. It carved out a new brand space, one that represents the intersection between the bargain hunting coupon clippers and the connectivity afforded by social media, creating a sensation that yielded $500 million in its second year as a start-up and attracted Starbucks' Howard Shultz as a board member.

When a company circles in on a new spot in the competitive space, it can own that "wrist" for a time by redefining rules and establishing new standards. Whether it's Amazon creating ease through one-click online shopping, JP Morgan revolutionizing banking with check scanning on a mobile phone, or Pei Wei raising the bar for upscale quick casual, a myriad of examples exist of companies that have looked at their world through a new lens and taken a new piece of the pie.

5. Spin the globe: the reality is that every brand is touched by global dynamics

Even locally based companies are affected by global economic factors and dynamics. When we hire talent, we're forced to look at global competitors to help us set standards for wages and work expectations. When we look for suppliers, we add global comparisons into our equations. And whether we're a large, global enterprise or a small, local company, at some point we'll experience the impact of global dynamics.

When it comes to our brands, we have to take off our local "blinders" and expand our fields of vision to include issues that are important to our customers, whether those issues are worldwide poverty (aka Toms Shoes) or good, fast Asian fare worldwide (like P.F. Chang's). Our brands today have to transcend the feeling of "domestic" and move into global.

Find Your Next in Secret Sauce

1. Build a dynamic dashboard. Forget fixed values, and try "feeds." Follow the feedback in your market. Cobble it together.
2. Watch for early signs—customer satisfaction and wow factors— that your brand is on the decline. Coke and Pepsi both came out of the cola chutes on equal footing, but Pepsi lost ground. Does your brand need a turnaround?
3. Capture feedback on what's missing: who's entered the market and captured market share? Is there a "wrist" you can own, a new place in the landscape that you can claim for your brand?
4. Ride the wave. Morph to the place where your brand's value is heading. G.E. developed ecomagination and installed it into its manufacturing redesign. Johnson & Johnson embraced the growing Asian market.
   What's in the future for you?
5. Plug in to the impact of Web 2.0 on brand loyalty. Listen for the creativity in the social media conversations about your brand. Take the customer's input on the whole industry. Are there patterns emerging where you could build new impact?

Chapter

9

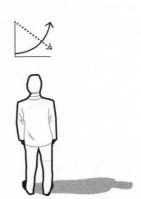

Trendability

Stay Ahead of the Unpredictable: The Game of Trending and Adaptation

Your Self-Diagnostic of Trendability

1. You can't predict what will come next. But you must respond when it does come. Are you equipped to adapt, refocus, and thrive in the new competitive landscape?
2. The last era was about models and forecasting. Today's era is about foresight. Are old methods blinding you to upcoming shifts in the business ecosystem?
3. Once you start looking ahead, you might be looking in the wrong direction. Will you miss the next big thing that could turn your business upside down?
4. The new world is full of trends. Which ones should you follow?
5. Some companies have figured it out. What exactly have they figured out and what can you learn from them?

In a great commercial about Las Vegas, a couple somewhere in the Midwestern United States rides around their neighborhood in a car equipped with a GPS (global positioning service). The device speaks in a lovely, soothing voice and provides step-by-step instructions to the disoriented husband and wife, who obligingly follow every command to the letter. As the car starts to travel toward unfamiliar territory, the husband turns to his wife, shrugs his shoulders with a look that says, "I'm just following instructions," and continues to drive very, very far from home—all the way to Las Vegas. But that wasn't the original game plan.

Did the couple actually want to go to Vegas? That question is up for debate. The point of the story is not where they were going—Vegas or not—but who was in control: the couple or the GPS. The GPS made it easy for them to go into "auto pilot" mode, and yield to someone else's plan for their future. Not necessarily a good idea.

Too many companies focus only on those goals in plain view, initiatives that are sometimes based on expired plans and projections that need to be revisited. That approach can steer a company in the wrong, untrendable direction. Companies that don't have a dynamic, active way to incorporate trends into their DNA will get just as turned around as the couple heading to Vegas: falling into an old pattern of obeying irrelevant strategic guidance systems, blind to indications that they might be off

Stacked: iPad It Your Way

Look at the founders of BJ's Stacked as a great example of trendability. Stacked is a new burger restaurant that takes the "have it your way" philosophy to the next level by enabling guests to order their burgers on iPads. The ordering system accounts for every detail (with great photos): customizable buns, sauce, toppings, and add-ons. At the end of the meal, there's even a card-swipe system for payment.

The iPad is changing retail. With it, a hair stylist can instantly look up images of a short pixie cut, or the latest moviestar do. And restaurants can take the Burger King motto full tilt and do it the customer's way. While ordering, customers of BJ's Stacked make it custom watching on the iPad what they're actually getting ready to eat. As they add and subtract, the price goes up and down accordingly, as does the photographed hamburger. It's a model fitting for the times: one of the customer and one of information.

course or losing ground to competition just as high-impact trends enter the picture.

Some companies get it right and seize those trendable opportunities. By spotting a trend and integrating that trend into their business, such organizations readily distinguish themselves.

Many companies have tried to integrate trend watching into their overall business strategy, without much success, achieving only a sense of hopelessness rather than the broader view they were after. What they need are new antennae that will enable them to sense the trends on the horizon—with enough time to respond quickly and effectively to any embedded opportunities. And, while mathematicians and models aren't entirely to blame, we mourn the comfort of the past, a bygone era when trends gave fair warning and our predictive models gave us time to cycle through before going obsolete. Trend watching can paralyze us, or push us to get with the new program of organizational responsiveness: trendability.

Seeing the future used to be a matter of trying to predict it. Think back as early as the 1920s. In 1929, just before the U.S. stock market crash, G.V. Cox published an article analyzing the relationship between economic forecasts and financial outcomes (Rotheli, 2007). Forecasters like Moody's and Babson's established credibility with the idea that businesses needed to integrate forecasting models into their strategic planning; the power of financial modeling and technical research as a driver for business growth took hold. John Moody, founder of a ratings service for public market securities, came to the discipline of data analysis during the months—and actualities—following the Great Depression: at the time, he could correlate his list of highly rated bonds with low default rates (Moody, n.d.).

And so, companies like Sears, DuPont, and Procter & Gamble incorporated predictive science into their corporate strategies, reaping the benefits of growth for decades. Procter & Gamble was an early leader in scientific research, establishing a privately funded research and development facility in 1890. The company applied that same research focus to customer research, pioneering methods for consumer research in the 1920s that would drive the creation of brands like Pampers, the first broadly available disposable diapers, and Tide laundry detergent. Procter & Gamble's research mindset also made it a leader in forecast modeling and demand planning.

But even Procter & Gamble, with its century-plus track record for driving innovation based on forecasts, research, and predictive modeling, balances its science with the art of observation by emphasizing consumer and climatic adaptation. Dick Clark, demand planning global process leader for Procter & Gamble, cautioned against overdependence on forecasting (in his case, demand planning) as a panacea to uncovering the right future path for a company: "Demand planning does not have a crystal ball. There is no silver bullet. There is not one work process, one training class, one software solution, or one method of collecting assumptions that will eliminate forecast bias. We must recognize that it is macro assumptions, management over-rides, and other big decisions that lead to forecast bias. In the current economic environment we are seeing unexpected changes in market size and dramatic changes in customer, shopper, and consumer behavior. We need to better understand the relationship between market size, shipments, and share through collaboration with business partners" (Clark, 2009).

Other recent, dramatic lessons reinforce the need to temper our reliance on brilliant mathematical modeling with a strong dose of observation and readjustment to change. Models alone, no matter how robust, are not enough to provide 100 percent certainty about what to do next. Not only can models be fraught with bias, but they can miss a critical element, as in the case of the string of events leading to the recent financial meltdown (Patterson, 2010).

Predictions are powerful only when they turn out to be true. However, betting the future of a company on only one part of a three-part equation consisting of models, observations, and adaptation can be very risky, as when Wall Street "began to explode in spectacular fashion beginning in August 2007" despite brilliant theoretical models (Patterson, 2010, p. 12). It's time to put forecasts and predictions into a more balanced place in the overall equation of trendability, and embrace a new orthodoxy that puts numbers where they belong. It's time to take our analytics off "auto pilot" and open our eyes to what will allow us to respond to a future we can't see.

Gaining competitive advantage in today's ever-changing business world is an entirely different matter than it used to be. So how do you apply trendability to gain competitive advantage? For one thing, avoid the traps. You can get so focused on estimates and predictions that you're ill equipped to handle change, whether it's still to come or right in front of you. And even then, when you're facing first-time downturns, unanticipated tightening of margins, and new competition, it can get really complicated.

You don't want to be Palm with a singular attention on forecasts of Blackberry's sales just as Apple releases the iPhone and Google announces the Droid (Svensson, 2010).

The focus on forecasting can fog the windshield, and more importantly, the road ahead. Imagine being deeply engaged in economic analysis, too intent on the future direction of your current competitors when news as dramatic as the iPad suddenly appears on the radar screen. And picture your company doing nothing—not responding at all—simply because your organization has never practiced, or learned, the art of rapid adaptation. When you do see a change coming, and try to swerve, you won't be able to get out of the way.

Let's look at Warren Buffett, the third richest person in the world, a man who continually reaps great rewards from his ability to predict. Even Buffett has to admit defeat sometimes, regardless of a long history of success in foreseeing what looks unknowable to the rest of us. He has, in fact, come face-to-face with the limitations of his predictive decision models. They don't always work.

Warren Buffett made a strong case for shifting to rapid adaptation when he moved to reverse an investment decision. In a "very expensive business fiasco entirely of his own making," Buffett thought a credit card for his car insurance company, Geico, made famous by the British-accented gecko, was a great idea. In actuality, the prediction was based on sound assumptions: "I reasoned that Geico policyholders were likely to be good credit risks, and, assuming we offered an attractive card, they would likely favor us with their business" (Barr, 2010). He started using it, and even pushed his shareholders to join him. It wasn't until Geico lost $6.3 million pretax on cards that he finally woke up. The loss kept going, totaling $44 million when it sold, a $98 million portfolio of card receivables, at 55 cents on the dollar.

Throw out those old paradigms from business school, in which all graphs are on the money, economic models play out exactly as predicted, and well-researched assumptions automatically turn into profits. We can no longer avoid being adaptive—if we want to survive. Whether driven by a need to adjust to a world where our models and predictions don't capture new realities, or motivated by a call to respond quickly to new circumstances, we all need to master the art of sensing change early and the discipline of adapting to it.

The cost of failing on either count can be extinction. What keeps your company from adapting? Or not adapting?

The New Realities of Trendability

1. Early sighting of trends can lead to market dominance.
2. Responsiveness to change is the new killer application in the corporate arsenal.
3. Technological innovation can threaten the touch of personal service, but it can also create opportunity.
4. Societal shifts can change consumer preferences and open up the market to new competition. That can mean new product and service offerings. Watch for them.
5. The signs of our future already exist in adjacent and even unrelated industries, and provide insights into consumer preferences.

---

1. Early sighting of trends can lead to market leadership

Apple is the gold standard of trendability. The innovator got its start in computers with its attack of an untapped market—schools—by donating a free computer to every school in California in 1983. It was a move to get computers in the classroom and it worked (Uston, 1983). Apple saw that personal computers dominated the workplace at the time, but not the schools. That territory was ripe for a new brand—the

K through 12 emerging generation of computer users. The company's bet was right, and Apple computers succeeded in establishing brand loyalty in a new niche. Apples became the "not the office PC" brand for millions of students, paving the way for Apple to innovate with products that would capture the imagination of a customer base that valued "edgy."

In 2009, Autodesk experienced the same sort of revelation about a new segment of customers that was currently going underserved by its offerings—the consumer market. The historically professional-focused company recognized it could either continue to go down the path it was on, providing software and solutions to professionally trained designers and engineers, or it could venture into new territory. And so it did.

Autodesk: Beyond Connecting the Dots

Autodesk had always believed that the innovation design tools it provided professionals—architects, engineers, industrial designers, game developers, special effects artists, and the like—had the potential to unlock creativity in all genres of consumers—kids, hobbyists, and do-it-yourselfers—but only if the design software leader could find the right combination of user interaction and capability.

Several times over the course of the company's nearly 30-year history, Autodesk tried unsuccessfully to find the right combination of user interface and functionality that would allow non-professionals to use its world leading design and creativity software. The problem lay in how consumers create. While professionals were willing to use the mouse and keyboard for design, many consumers just couldn't get comfortable expressing their creativity with that type of interface. It was simply too limiting and foreign to them when it came to creating art and design. Also, professionals were willing to forgo photo-realistic designs until the end of the creative process. Traditional Autodesk customers were comfortable designing and creating with line drawings and wire frame representations that used simple color-shaded objects until the end design was nearly complete. But consumers wanted photo-realism all the time. For example, they wanted to drag and drop refrigerators and countertops with full fidelity into kitchen designs so they could see exactly what the final kitchen (and all of its details) would look like.

(Contnued on next page)

(Contnued from previous page)

Years ago, Autodesk launched products like "Kitchen, Bath, Deck" and "Picture This Home," but none of them were successful because they all struggled to adequately address the design needs and nuances of the non-professional. It was not until the arrival of touch screens and cloud computing—the perfect answers to the challenges that had plagued Autodesk's consumer aspirations for nearly three decades—that Autodesk would find success with consumers.

Mobile computing allowed Autodesk to introduce the finger as a creative input device (no longer just the mouse and keyboard). All of a sudden the company could give a new customer segment, from kids to creative artists, the ability to draw with their hands. Children around the world have known the power and enjoyment of finger painting and sculpting clay (or Play-Doh) with their hands. Autodesk tapped into this native form of creation through its Sketchbook and 123D Sculpt apps. Today there are more than 5 million users of Autodesk Sketchbook on Apple iOS and Google Android-powered phones and tablets, with more and more joining every day. Autodesk 123D Sculpt, released in mid-2011, is for those that want to use "virtual clay" on mobile devices to create anything they can imagine—but without getting their hands dirty.

Cloud computing allowed Autodesk to solve the second problem: delivering photo-realistic 3D imagery to consumers in real time. With computing power in the cloud able to quickly process images, Autodesk was finally able to create an application that allowed anyone to play "what-if" with the home furnishings, fixtures, flooring, and designs of their choice. Called "Autodesk Homestyler," the cloud-based application currently averages more than 660,000 unique visitors every three months. It lets anyone in the world change the design of their living space and experience a photo-realistic rendering of a new virtual space with a simple drag and drop.

Chris Bradshaw, Autodesk's chief marketing officer, explains that Autodesk's culture had to stretch and its people had to learn new ways to get their arms around the mindset of their new customers. "We're still learning the full potential in this new world of consumer design. So far 25 million people have joined our design community and embraced our brand as users of our portfolio of consumer applications. We have more products on the horizon, and we continue to learn more about how to develop software differently for this new type of Autodesk customer. For a company like ours, so dedicated to creativity, it's stimulating to be in this new world of discovery."

The emergence of a trend is obvious in retrospect. We can all look back and see the logic behind entering a new consumer space the way Apple and Autodesk did. But it takes a special type of corporate discipline to train ourselves in trendability, and to take the time needed to recognize the early rumblings in our competitive landscape that are actually seismic shifts, and when a new market, a new customer segment—a new opportunity—is beginning to emerge.

2. Responsiveness to change is the new killer application in the corporate arsenal

We've all had moments when we've hit a wall: despite what the numbers say, sales are flat, even though expectations for revenue growth remain high. We're executing well, but the market isn't responding the way it used to. We feel a sense of urgency to shift gears, but we're not sure what to do next. But we need to respond.

Polaroid was a company too sluggish to enter the world of digital photography. Kodak wasn't. Kodak transformed the preservation of memories by embracing the potential of digital technology and developing services like Easy Share, while Polaroid descended into bankruptcy (Franklin, 2011). Kodak proved that learning to read the signs of change (even those that don't appear in traditional forecasts) and creating an organization geared to rapid adaptation is the only way to survive in the new world of fast-paced competition.

Despite our best intentions, Polaroid isn't the only example of extinction thinking to litter the floors of business cemeteries. Recall every Napster moment (Boutin, 2010), when the revolving door of opportunity swings one company out on the street, allowing other competitors to race in and grab those customers, sometimes overnight.

One insurance company I worked with almost experienced its own such Napster moment, a near-miss that could have lost the company its entire foothold in the market in a flash. The company's management was trying to drive sales and doing well, or so it thought. Having made an aggressive sales push to recruit younger agents, the company had just begun to gain traction with younger, more affluent clients. Pleased with what the team saw as a lock on relationships with new clients, it got together to strategize for a future based on a well-executed game plan. Management agreed its strongest competitive card was its loyal base of agents, a group that had been nurtured for more than two decades and brought 80 percent of its new business to the company.

But the story doesn't end there. One of the executives, Chris, was paying closer attention to trends than the rest of the team was. He foresaw what might have resulted in a possible chink in their corporate armor—new capabilities in technology that could make agents less central to insurance sales. The threat on the horizon was real: the company's perceived invincible hold on customer loyalty was based on long-standing interpersonal relationships with agents in the field—and nothing else. Could it survive if agents were no longer needed? And if the next batch of customers didn't want great personal service, what did they want?

Chris didn't know, but he could find out. He called an expert, an executive from a large credit card company that had already made the shift to online processing, and asked him to share his observations on this new market: 25- to 40-year-olds with a lot of money to spend. Tucked within the speaker's technical presentation on the intricacies of billing, online payment patterns, and demographics was a "by the way," a sneak attack Chris's people weren't ready for. More specifically, it went like this: "Every month, we send credit card bills out to our customer base, which represent approximately 60 percent of the total market. We've actually considered a slight modification to the bill where we add a small check box that says, 'Check here if you'd like to buy life insurance.' That way, we could simply add the cost of a basic policy to their monthly bill. With one relatively simple addition to our standard form, we could enter the insurance business overnight." The idea was a good one—for the credit card company—and a bad one—for Chris's firm. The impact of that single idea could have transformed the credit card company into the trusted source of a new add-on service. Over time, it might have been able to translate the customer loyalty it had from its current credit card business to an expanded offering of life insurance.

The good news for the insurance company was that the credit card company decided not to add that check box. But if it had, it would have entered Chris's world as a competitor with unrivaled access to the same customer base and an easier billing infrastructure. And Chris's company would have lost its lock on its customers.

But his original concern—the move toward online—was still an issue. What Chris saw early on was that the credit card industry was moving rapidly toward a new standard for cementing customer loyalty, one Chris believed would soon become the prime advantage of all financial services companies. Chris believed that the age of the agent as king was over and the age of the permission-based electronic customer relationship had begun. And he was right.

It was only a matter of time before the wake-up call would play out in powerful new ways, because within a few years companies like eSurance and eTrade would hit the scene and prove the point Chris was making:that he needed to be extremely skilled at the science of responding to wake-up calls.

And once that trend or wake-up call is upon you, you must respond quickly and nimbly.

Norma Kamali: High Fashion at a Lower Price

Norma Kamali responded to the downturn in the economy, and quickly, knowing it would mean less people buying wardrobes at four-figure price tags. As a trend-sensitive, high-end, exclusive clothing designer, she offered ready-to-wear versions of her new line, with a runway show devoted to her theme: The Democratization of Fashion. It was sponsored by Mercedes Benz, another brand seeking entry into the middle-income market as a way to hedge its bets at a time when the purchase of luxury goods was off-trend. In the downscale version of a traditional Fashion Week event, the Norma Kamali models became walking ads, holding signs about the economical eBay line for only $250 and under and some pieces for just $35 and under for purchase at Wal-Mart (Ho, 2009).

The Norma Kamali brand's rapid response to a new economic reality sets the pace for the moment we all confront when change is upon us—the lesson for all of us is to embrace the new facts and nimbly devise a shift in course that seizes the up-side of the moment.

3. Technological innovation can threaten the touch of personal service

In 2001, I stood in the doorway of Charlie Trotter's restaurant in Chicago and watched as the sommelier recited the finer points of the restaurant's well-crafted wine list. She prided herself on her knowledge of vintage, employing her virtually photographic memory and impeccably trained palate. The value of her education and her immersion in the subtleties of her product placed her in a class of her own, and that knowledge distinguished the restaurant as a premium brand, worth the premium prices it charged. But something would threaten the sommelier and her recall for wine. Within two years, the value of being the go-to person for the wine cellar's contents would be challenged by the invention of a new generation of technology that combined hand-held scanners—like the kind rental car companies used—with a rich database of wine facts and data.

Now we wouldn't need that elegant lady behind the bar, dressed in black and white, so classic she's almost forgettable. The latest in gadgetry, an ultra-high resolution mass spectrometer can tell you everything you need to know, another example of technology taking the place of a living person (Keim, 2009). Would the sommelier become a dinosaur? Would the restaurant have to look for new points of distinction?

Fast forward to 2009, and you'll see that, yes, the old school version of a sommelier might be on the verge of becoming obsolete, but at least SD26 restaurant in New York City took a new tack with the challenge: an electronic wine list. During a recent visit to SD26 in New York City, I experienced Rosie the Robot from the *Jetsons*—technology-enhanced savvy in the form of the sommelier, who was armed with a handheld device called the Smartcellar. The Smartcellar provided complete details on every bottle of wine on their list, including vineyard, region of origin, price, producer, and its availability in the restaurant that day. It provided me with the best of both worlds: high-tech, encyclopedic knowledge plus the personal engagement of a sommelier.

4. Societal shifts can change consumer preferences and open up the market to new competition

Think back to the 1990s, or even what came before that pivotal decade, when *Mad Men* wasn't just a sitcom, but a reality. We were bumping up against repercussions of an emerging trend: women in the workforce, and more specifically, the professional arena. But this widespread demographic movement manifested in ways we might not have predicted, such as food. Women have a role men don't have—as mothers—and as a result they don't think about food the same way men do, even fathers.

The restaurants felt the impact of these changes first. A decade ago, sales of ready-made food items at grocery stores like gourmet potato salad and fresh cut fruit (i.e., Houston-based Rice Epicurean Markets) were on the rise as restaurant check averages were on the decline. The market was shifting. Customers—female customers—were looking for a new product that restaurants couldn't provide: chef-prepared food they could serve at home. Women's entry into the workplace and a disposable income had led to a second wave of the trend. With children at home and less time to cook, the convenience of the restaurant was, in this second wave, being trumped by the desire to spend less time making dinner, and still give their families that feeling of a homemade meal. Some restaurant companies, like Brinker's EatZi's, led the way for a new category: home meal replacements. Others let the grocery stores capitalize on that trend, paving the way for the likes of Whole Foods Markets and Trader Joe's.

Women keep looking for that perfect replacement to cooking, and perhaps they still haven't found it. It didn't stop at ready-made meals, though, and the industry will continue to evolve. Author Michael Pollan, author of *In Defense of Food*, notes the decline in cooking and the rise of home-meal replacements: "That decline has several causes: women working outside the home; food companies persuading us to let them do the cooking; and advances in technology that made it easier for them to do so. Cooking is no longer obligatory, and for many people, women especially, that has been a blessing" (Thomas, 2009). It seems only logical that another type of food delivery service would be on the rise, the personal chef, with around 9,000 personal chef businesses in the United States, an increase from 1,000 a decade ago (Black, 2005).

Walgreens: A Walk Down a Different Aisle

Walgreens has stepped up to the prepared foods trend, adding fresh fruits and vegetables and other ready-to-eat foods to every store in its network. Now it won't just be competing with other drugstores, but grocery stores as well, and being smaller and more convenient, it could easily be that "stop on your way home" place for something for dinner. Jumping in and out of a drugstore is infinitely easier than jumping in and out of a store like Target or Kroger. And others have stepped in to expand the landscape of when and how we eat—Jack-in-the-Box staked a claim on the late-night meal and food trucks entered the scene as a more flexible option that could roam from location to location to draw new crowds.

Restaurants and their spin-offs experienced a societal shift that didn't go away, as people moved (and continue moving) toward the same product serviced in a different way, and women began leading the way toward new options, new choices, and a huge challenge for the restaurant industry to tackle.

Food industry winners like Whole Foods watched trends and morphed their offerings as preferences shifted.

5. The signs of the future are here today

It isn't all bad news. Some changes could mean more revenue sources, but you have to be paying attention or they might pass you by. Let broadcast radio serve as an example of what happens when tunnel vision keeps you in the dark until it's too late.

With new alternatives for listening to music, satellite radio like Sirius/XM and Pandora stand out as examples of a climate change with permanent chilling effects throughout the entire world of traditional

broadcast radio. Even the forecast for 2010 looked grim, with an antici-
pated increase from 12 million units of satellite radio devices in 2005 to
55 million units in 2010, a rate of increase of 35 percent (Sennitt, 2009).
And it has meant a huge loss of ad revenue for traditional broadcast
radio stations.

Clues can be found in reports that reconfigure the component parts
of traditional broadcast radio and reassemble them into a new category:
"By 2008, online advertising spending in the United States is expected to
surpass radio advertising spending. While not signaling the death of
radio, it is an indication that radio is being subsumed into a broader sector called 'audio' Internet and satellite radio, podcasting, high-definition
radio and mobile audio services are all revolutionizing a radio industry
that remained virtually unchanged for a century" (Macklin, 2008).

Radio's Tidal Shift

Jim Kerr, from Triton Digital Media, summarizes
his view of the trends that will affect the future of
radio, with five points that are all aimed to
encourage innovation in the field (2011):
1. Radio is no longer local.
2. We are listening in the car by way of apps.
3. Customizability of radio listening.
4. iTunes are on the road.
5. Listener statistics are right now.
Customers are changing, and so is the way they listen to music.
And that means radio, which "remained virtually unchanged for a
century," is looking behind them, in the rearview mirror, at a
wake-up call.

Every business can learn from broadcast radio's evolution from
the world of disc jockeys to the world of podcasters, bloggers, and
mobile messaging. "Yesterday's podcasting experts are today's social
media gurus ... Yet, podcasting has never been more popular, never
touched as many lives, and never made as much money as it does
today" (Webster, 2010).

And, certainly there's reason to be optimistic that the reshuffling will lead to overall growth in the market for mobile audio. Using traditional television as an analog, we can recall forecasts that traditional television would be dead (Arrington, 2006), a prediction that turned out not to be true after all. We can trace the march through time from three television networks, to cable, to ubiquitous programming available on screen, to TiVo, Netflix, mobile devices, and the Internet, with repurposed, crowd-produced videos that have viral lives through YouTube and Vimeo (Worden, 2011); and at the same time we can learn about how to re-categorize our assumptions, rethink our forecasts, and uncover new paths for business growth.

Find Your Next in Trendability   🔍 find

1. Take the pulse of trends within your own industry, and the back-of-the-envelope view of your ecosystem. If you're in an industry in a closed system, in which the loss of sales in one area points to a rise of sales elsewhere in the system (the waterbed effect), ask yourself whose sales are rising as your sales are falling.
2. Step back from the day-to-day and reflect on early signs of change. Start with three core categories—technology, societal shifts and economic change, and customer experience—and create an informal dashboard to map trends that could translate into competitive threats.
3. Diagram the events that fundamentally change consumer expectations. Some companies enact a broad sweeping innovation in basic rules and everyone has to follow suit.
4. Assemble ratings lists in the core categories that can be the levers for competitive advantage: product innovation (Fast Company, 2010), talent and culture, processes (Baldrige, 2010), brand differentiation (Millward Brown, 2010), and customer loyalty (J.D. Power & Associates, 2011). Piece together a profile of what it will take to stay ahead of the future in each of the core categories.
5. Look inside your industry and look across industries. Compile the trends that will hit you hardest based on what you observe today and embed those insights into a trend-watching dashboard that you can focus on today to stay ahead of the future.

Part
3
CASE STUDIES

Case Studies

In school, history always looked neat and clean. Teachers made it that way, because they knew exactly how it all turned out. And because they knew, they were at a great advantage, and able to tell us every war story however they wanted to, using whichever perspective would best fit the conclusion they were leading us toward, using what was a very complex set of facts. All of the messiness and ambiguity of the moment could be filtered through the lens of "we know who wins." Business case studies—or the real stories of victory and defeat—can feel like that too. Every example is a tidied-up model of how things work in a real life company and isan efficientand entertaining way to teach the basic principles of business to students.

Unfortunately, when we enter the real world of business, things get a lot messier than they ever appeared in the retelling. It's frustrating to promise growth and then lay out the tools we have in front of us and wonder what to do with them, and how to get to our next: "How can I optimize the right processes to get our division's sales to grow?" "Where will my next customers come from now that we seem to have fully penetrated our current markets?" "How can I create a culture that will help me stay ahead of the trends I see?"

The real stories that follow are actually case studies that are universally applicable and laid out to demonstrate how things really work in business. No one was prompted to shoehorn his or her realities into a clean model. Because it isn't possible. The explanations of how each company moved from point A to point B are not organized or linear, or exactly the same. What they all have in common, however, is a sense of passion on the part of the leaders to figure things out and find the best approach to growing their companies.

Each of the companies in the following case studies illustrate some demonstration the *Find Your Next* steps:

1. Sort,
2. Match your genome,
3. Hybridize using cross-industry sources of inspiration, and finally,
4. Adapt and thrive.

But the hows of each approach will inspire you because you won't know exactly which formula will work to move your needle forward.

The following stories are where the six genomic elements combine in real ways to stimulate new visions for the future in organizations ranging from popular restaurants concepts (P.F. Chang's) to visionary technology companies (EMC Corporation). You'll understand the commitment of this diversified range of leaders who share common traits, and attributes that many business leaders possess: a genuine love of their industry, their company, their teams, their customers, their processes, their stakeholders; a faith and belief in what they sell; and a dogged sense of pursuit of their most powerful, highest-impact next.

P.F. CHANG'S CHINA BISTRO

Laying a Golden Egg: P.F. Chang's China Bistro Feeds Its Nexts

One sunny afternoon in 1999, Bert Vivian and Rick Federico were having a hard time hearing each other. They were seated at a table by the front door of their company's hot concept restaurant, P.F. Chang's China Bistro, which, at the time, was one of 29 units in a fast-growing chain. Located just a stone's throw from the corporate headquarters in Scottsdale, Arizona, this particular "bistro" was the original, the very place where the goose was born, the P.F. Chang's brand that would come to lay golden eggs. The brand's reputation had already been established as a highly successful restaurant concept, having cracked the code on delivering Asian food in the context of a chain. And, it would soon set new industry standards in the concept of brand extension by redefining the bistro.

Ever since joining the P.F. Chang's team in 1996, Rick and Bert had both made it a habit to take time out from spreadsheets, menu testing, PowerPoint presentations, board responsibilities, and the office atmosphere, in an effort to simply immerse themselves in what fueled their passions: the restaurant's vibe. That day in 1999, Bert and Rick did what they do best they followed their instincts, went after an idea, and got lucky. As a result, the concept evolved, and so did the restaurant.

That day, the waiting crowd was animated and lively, and loud. But it was more than just background noise that caught Rick's and Bert's attention. More distracting was where their table was situated: right by the to-go line. Rick stopped mid-conversation and watched the people as they entered—dressed in baseball caps and flip flops—struggling to push their way through tables of diners to get to the cash register and their plastic bag of food. He noticed the irony of the inconvenience: saving time was exactly the reason these people were getting their food to go, but there was no direct pathway to pick up their food and get out. As Rick tells it, "I'd always hated our carry-out experience. It took the back burner to what we were in business to do in the bistro environment:

offer a special dine-in experience, with every step of the way well-orchestrated to wow the guests."

Starting with this powerful observation of the guest collision, Rick knew they'd inadvertently created a lose/lose situation. What he didn't know yet was that by moving the back burner problem of the under-served casual guest to the front burner, he would help build a new concept that would extend the core "bistro" brand by adding its own, unique appeal.

Rick came to what would be a very pivotal revelation. Not only was P.F. Chang's China Bistro not doing justice to its food by offering carry-out as an afterthought to the dine-in experience, but it also wasn't accommodating an emerging and increasingly important need of its guests: to experience meals without having to dine in. The restaurant's current waiting area was a one-size-fits-none and Rick saw that it was time to make a change.

That very well-informed hunch became a catalyst for a new idea, and was a flash of insight that would translate into the next stage of the brand's development: expanding the core restaurant concept to move P.F. Chang's China Bistro to a new level.

As Rick watched the takeout crowd interrupt the waiting diners, he committed himself and the company to a new approach: "fast casual." He realized that the trend of eating on the run was here to stay, which meant take-out was no longer the exception to the rule, but a new rule.

For Rick Federico of P.F. Chang's, that revelation was built on industry trend statistics that he, Bert, the management team, and the board had already been contemplating. Together they witnessed the emergence of the "fast casual" as its own segment, one with strong appeal as a possible next formulation for P.F. Chang's China Bistro. At the time, Panera and Chipotle were already setting the pace. By 1999, Panera had sold off its Au Bon Pain units to "bet the future of the company on Panera Bread (Panera Company n.d.)," and enjoyed head-turning success, while Chipotle had been enjoying double-digit growth for a few years running (Gogoi, 23).

Despite the compelling evidence of a growing trend in "fast casual," Rick had to see it in action with his own eyes before he could fully buy in to the idea. But now that he had, he'd crossed the Rubicon, which allowed him to finally shift into commitment mode.

Pei Wei Makes New Rules: Next Pace-Setting

Not long after that lunch, Rick and Bert put together a team to right the wrongs they'd seen in the bistro, and disentangle what had become a collision of diversified guest preferences. They needed to maintain the integrity of their bistro's food in a service model that accommodated every kind of diner: the dad on his way home from coaching soccer, the young professional taking a dinner break between deadlines, and the laid-back couple looking to go out for a casual meal in the middle of the afternoon. Pei Wei was the answer. The new restaurant was embraced as P.F. Chang's "next" and set new standards in what did in fact turn out to be a growing category in the restaurant industry.

Now, in 2011, years later, the restaurant industry still feels the "glow" of the "fast-casual" trend and its growth trajectory, despite the recent lagging economy.

"Fast-casual menus tout premium ingredients and taste profiles, which translate to consumers as a healthier buy, worth the slightly higher price point" (Davis, 2011).

Today, Pei Wei Asian Diner has 168 units, P.F. Chang's operates more than 200 bistros, and the company is on solid ground for the future. But, with the company's original concepts starting to mature, what will be P.F. Chang's China Bistro's "next"? What is core to P.F. Chang's corporate DNA that might be nurtured into a sustainable new direction for the company?

The Next Next: Plant Your Seeds

Mike Welborn, president of global brand development for P.F. Chang's, takes a refreshing look at corporate nexts. His motto? "Build them before you need them." But picking a new direction can be daunting, even for a company with a great record for business savvy and good fortune.

"I was on the P.F. Chang's board from the early years, and brought a non-industry lens to the topic of growth. I'd studied strategy with

Michael Porter [Porter, 1980] in business school and had worked under great leaders like Jamie Dimon, then CEO of Bank One [now part of JP Morgan Chase], when I was in the financial services industry, but that didn't mean it would be simple to pick the right path to focus on for the post-Pei Wei era in P.F. Chang's development" (Welborn, 2011).

In the years since Welborn joined P.F. Chang's board as one of the original investors, he's been part of many discussions about extending the brand, and puts in practice Michael Porter's wisdom of thinking about strategy all the time. Strategy, as great companies like G.E. well know, is not a once a year exercise or a focus that comes from being painted into a corner, with no foreseeable attractive options on the horizon. According to Porter, "If you want to make a difference as a leader, you've got to make time for strategy . . . but setting strategy has become a little more complicated. In the old days, maybe 20 years ago, you could set a direction for your business, define a value proposition, and then lumber along pursuing that. Today, you still need to define how you're going to be distinctive. But we know that simply making that set of choices will not protect you unless you're constantly sucking in all of the available means to improve on your ability to deliver" (Hammonds, 2001).

When the reality hit Welborn and the rest of the board that it was time to grow P.F. Chang's, they experienced what Mike describes as "the difference between the way things look in the neat telling of a story after the fact, when you have the luxury of looking back on a series of steps, when all facts are known and the ending has played out, versus the heat of the moment when decisions are in your line of sight and you have to choose a path" (Welborn, 2011). They wanted to pick their next step wisely.

The company had suffered a couple of false starts as a result of moving toward new concepts that didn't pan out, for either de-focusing management from the company's core business or for their choice not being the right next. In one such example, Taneko Japanese Tavern opened with high hopes and good reviews in 2007. But, ultimately, Taneko didn't meet the expectations for brand extension; still, all was not lost, because the misfire did teach the organization what would constitute a better option, one that would better extend the brand and fill a missing need for its guests, once they put the tavern to bed.

According to Welborn, "it's not always a crisis, or pressure for double-digit growth, or a threat from the market that drives an organization's pursuit of its next. In our case, we had the luxury of healthy cash flows and a core brand with scalability, but we still felt the pull of possibility." By understanding that Taneko wasn't the right move, the organization could discuss growth more strategically. And Welborn thought back to a lesson he'd learned from Jamie Dimon, his former colleague at Bank One: "Let's get better at what we do today, and from there we'll land on a new direction."

Welborn believed it was essential to build on the core of the brand's magic: quality ingredients, great flavors, and a dining experience that turned guests into regulars. What would P.F. Chang's do for its encore?

P.F. Chang's Entrée into Cross-Industry Grafting: P.F. Chang's Goes Global and Frozen

If ten restaurant executives had gotten together in early 2009 and looked at the cards in the P.F. Chang's deck to make a bet on the future, chances are good that only one of them could have predicted the unique path the company would eventually follow. The team took what it did best and added a very key ingredient—cross-fertilization of contacts and expertise—to combine its own DNA and core strengths with the DNA of other companies that had successfully migrated their wares into the grocery store and across the world.

Lane Cardwell, president of P.F. Chang's China Bistro, brings to the job years of developmental and expansion experience from dozens of restaurant concepts like Steak & Ale and Chili's. He explains that sometimes being opportunistic is the most strategic approach. "Outside of a few big players like Darden and McDonald's, the restaurant industry isn't known for using theoretical trends and complex forecasting equations as the drivers of strategic direction. There's always been an element of serendipity and entrepreneurial talent that has led to successful expansions of some of the most successful concepts."

P.F. Chang's looked at its options for expansion and decided to build on the strengths of what it already did well—crave-able food with an

Asian flavor profile. The company's management expanded the brandby matching a core aspect of what italready possessed with a supplementary attribute that was only a slight departure.

In 2010, P.F. Chang's Home Menu hit the frozen food sections of grocery stores around the country as part of a licensing agreement with Unilever, the organizationthat successfully raised the bar in the frozen Italian food category with its Bertolli brand.

Some of the most familiar names in the restaurant world were moving into the grocer's freezer. P.F. Chang's, Burger King, and Jamba Juice all have recently licensed their names for new products to be sold in supermarkets. They join other high-profile restaurant chains like Marie Callender's, Starbucks, T.G.I. Friday's, and California Pizza Kitchen, which already have a substantial presence at the grocery store.

The most recent moves come at a time when many former frequent diners—scared off by the recession—are not returning to restaurants. "It's a search for new revenue streams," says Robin Lee Allen, executive editor at *Nation's Restaurant News*. "It's also a way to keep the brand top of mind." And a way for a chain to familiarize consumers with the brand before entering a new market with restaurants, says consultant Linda Lipsky.

New freezing technologies that are better at keeping items tasty have made the grocery's frozen-food aisle almost a "no lose" proposition, says Bob Garrison, editor of *Refrigerated & Frozen Foods*, a food trade magazine. Among the newer name brands at markets? P.F. Chang's. Even with new methods, frozen food isn't going to taste exactly like fresh, which makes Chang's deal with Unilever to offer eight frozen meals-for-two at $7.50 to $10 compelling. Chang's, with 172 restaurants in 42 states, has built its reputation on food that tastes great. The question is, can Unilever's new freezing technique do the trick?

"There were enormous discussions," says Mike Welborn, head of global brand development. "We can never do anything to tarnish the brand."

Based on its success with the Bertolli Italian frozen-food line, executives at Unilever convinced Chang's that they could work the same magic on its Asian entrees. But, says Gaston Vaneri, frozen foods marketing chief at Unilever, the company also needed Chang's name recognition and its know-how to make a frozen Asian food line a hit (Horovitz, 2010).

In the case of P.F. Chang's, the organization tapped into the expertise of its senior leaders and board of directors, aligning its talent in three arenas, specifically ones that could be grafted from companies they had worked with. Lane Cardwell, with experience as former CEO of Eatzi's (a Brinker concept), brought to P.F. Chang's an understanding of nontraditional distribution with the company's notion of chef-prepared meals for eating at home. P.F. Chang's board member James Shennan, also a board member of Starbucks, learned the ropes in global expansion through licensing partners. And, in addition, P.F. Chang's also leveraged one of Welborn's core strengths, scalability, having spearheaded significant growth for several financial services organizations.

Mike Welborn believed that "without expertise on the board in these mission-critical arenas, it would have been too risky to venture in either direction. With frozen food, you run the risk of eroding the public's perceptions of quality, and with a global move, you have to commit to a lot of education and training to teach the P.F. Chang's entire operational model. But, since we had sound introductions to people who had been down those paths with Brinker and Starbucks, we had an advantage."

In 2009, P.F. Chang's committed to launching two new strategic initiatives, with Mike Welborn leading the charge: a partnership with Unilever to create P.F. Chang's Home Menu, a frozen food line, and expansion into global distribution. The common denominator was a set of rules that they'd put together from having worked as a team for more than a decade. They would enter new territory only if they could meet two important initial criteria: one, any new initiative must build on the strength of the existing brand and culture, and two, any new leadership skills they needed would have to be grafted onto the organization by bringing in experts that already possessed the core capabilities. Fortunately, both of these new, soon-to-be-successful initiatives qualified.

The Devil's Details: P.F. Chang's Chants Mantras for "New"

Rick Federico elaborates on what was behind his decision to take on frozen food and going global, two initiatives that could seem a bit ambitious and possibly even schizophrenic to the outside analyst. He explains that both moves were born from what became new mantras for the company: one, start with love and passion and a sense that the new direction is a great fit with the company's capabilities and culture; two, ten seconds later, bring in expertise with the core skills to drive the initiatives hard; three then, make certain that the sexiness of the new doesn't erode the power of the existing brand, add stress to the existing staff, or in any way detract from the guest relationship; and four, finally, discipline the organization and the board to view each new initiative very objectively, using the mindset a banker would use to analyze each new investment. "We had to keep ourselves in check to make sure we hadn't just fallen in love with the new concepts" (Federico, 2011).

Not to mention the fact that Mike Welborn brought a unique portfolio of global deal making skills to the equation. In 2009, he transitioned from board member to the head of these new initiatives,

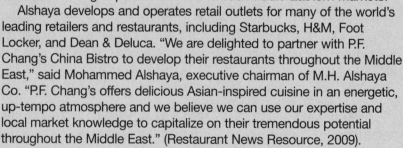

P.F. Chang's Expands to Mexico, the Philippines, and the Middle East

P.F. Chang's China Bistro, Inc., announced a development and license agreement with M.H. Alshaya, the Middle East's leading retailer, to develop 34 restaurants throughout the Middle East. "We looked at a number of potential development partners in the Middle East and Alshaya was the best," said Rick Federico. "We are excited to be working with them to extend our wonderful dining experience to numerous Middle Eastern markets."

Alshaya develops and operates retail outlets for many of the world's leading retailers and restaurants, including Starbucks, H&M, Foot Locker, and Dean & Deluca. "We are delighted to partner with P.F. Chang's China Bistro to develop their restaurants throughout the Middle East," said Mohammed Alshaya, executive chairman of M.H. Alshaya Co. "P.F. Chang's offers delicious Asian-inspired cuisine in an energetic, up-tempo atmosphere and we believe we can use our expertise and local market knowledge to capitalize on their tremendous potential throughout the Middle East." (Restaurant News Resource, 2009).

devotedly obeying each mantra. He and his colleagues drove hard to make a frozen meal relationship with Unilever a part of the P.F. Chang's culture. Before the deal was sealed, they staged a tasting of Bertolli's meals for the board to demonstrate how far the frozen food concept had evolved since the days of TV dinners. Before the launch, they made a video for employees featuring Philip Chiang, the creator of the brand's original flavor profile, explaining the care that had gone into translating the original vision for the food into a frozen meal format.

Welborn also followed the new rules when executing the vision for global expansion. According to colleague Lane Cardwell, Welborn brought exactly what the organization needed for global expansion: excellent understanding of the elements of business deals, international experience, and special talent for leveraging relationships for both the frozen food part of the business as well as the global partners that would drive expansion (Cardwell, 2011).

Both the Unilever deal and the global expansion path have turned out to be brilliant bets. Sales are up for Unilever. Food Ingredients First announced P.F. Chang's frozen meals a success, exceeding $50 million in turnover in the launch year (Food Ingredients First, 2011). According to Brand Channel, "Sales of P.F. Chang's China Bistro frozen retail items climbed 117 percent, to $14.5 million, in September, 2010, suggesting that it's possible for China Bistro to reach robust annual-ized sales of $117 million. And that's just in stores measured by Symphony IRI, which doesn't track sales via Wal-Mart" (Buss, 2010). The company reports good results outside the United States, as well. Its global division reported nearly $1 million in revenue for the third quarter of 2010, supported by frozen food and six international bistros (National Restaurant News, 2010). Nader Hallal, head of marketing for casual dining at MH Al Shaya, P.F. Chang's franchise partner, speaks well of the joint venture in May 2010: "We opened in Kuwait in December and it's been a phenomenal success. Even now, nearly six months later, the restaurant is as popular as on day one and people are still queuing up. So now we've got the appetite to really run with the brand and expand faster into new markets around the region" (Fernandez, 2010).

And, from the guest's perspective, the food remains a hit, getting high marks in both arenas. "It's all about the quality of the food and the

> ## P.F. Chang's Announces Loan Agreement with True Food Kitchen, a Fox Restaurant Concept
>
>
>
> P.F. Chang's China Bistro, Inc., announced an agreement to provide debt capital for the early-stage development of True Food Kitchen, a Fox Restaurant concept.
> The agreement provides for a $10 million loan to develop True Food Kitchen restaurants and can be converted by P.F. Chang's into a majority equity position in True Food Kitchen.
> "We have admired the work of Sam Fox and his team for quite some time," said Bert Vivian, co-CEO of P.F. Chang's China Bistro, Inc. "We are pleased to be able to provide capital for the growth of their newest creation" (Business Wire, 2009).

guest experience. We've put that front-and-center and constantly test the guest reaction to make sure that quality never slips even one iota. That's what has made all the difference, and we can never lose sight of that" (Federico, 2011).

Rick Federico also emphasizes the distinction between an "add-on" strategy, in which a new idea is simply pasted onto an organization, versus a true "graft," in which a new idea thrives. "It's all about our culture—that's our magic—how everything we do has to authentically reflect who we are. We've learned that nothing new can be done at the expense of our core culture. That seems like a subtle point, but it turns out to be everything. That's why we'd rather not pursue just any old distribution or brand expansion idea. We wait until the right thing comes along"(Federico, 2011).

No Time to Gather Moss: the P.F. Chang's Stone Rolls Toward Future Growth

You'd think that after such a stellar run on a growth path, P.F. Chang's could rest for a while. But, in 2011, the next chapter in P.F. Chang's rolling stone story, from bistro to fast casual to frozen at-home dining to global player, shows no signs of gathering moss. The "pull of possibility" that Mike Welborn described is at their doorstep again, as the company's management looks to what will yield even bigger numbers, upwards of

$250 million per trajectory, for new initiatives that round out the portfolio of the brand.

The winning formula for the company seems to be at work again as it puts its corporate toes into waters to test ideas like True Food Kitchen, a multiunit, privately-held concept that is the brainchild of Dr. Andrew Weil, healthy food guru, and Sam Fox, serial restaurant entrepreneur, spitting distance from P.F. Chang's Arizona headquarters.

In 2011, the company made an initial $10 million loan to True Food Kitchen, with an option to convert that loan to a majority equity position if certain conditions were met.

Why True Food Kitchen?

Rick Federico didn't hesitate to explain the lessons he learned from his P.F. Chang's experience. "We have a culture that thrives on a level of quality, integrity, and relationships that we can't compromise. So does True Food Kitchen. The passion for creating a guest experience of the highest caliber was one Sam Fox, of True Food Kitchen and Fox Restaurant Concepts, and I shared. That common bond was how our relationship began. Whatever we pick as our next next, together or separately, has to fit that bill.

If we go farther down the road with our relationship to offer people the best we can in food and service, we might have to bring in expertise to focus on whatever new portfolio we create together or whatever new direction we head toward. That's our rule for making sure our goose continues to lay its eggs as we feather the nest with new opportunities" (Federico, 2011).

Beyond "Next"

What's ahead for P.F. Chang's? How are they thinking beyond the next? Lane Cardwell is known by the restaurant industry as an expert on the next big thing, and he shares some of the whiz-bang ideas swimming in his head right now for the future of P.F. Chang's. He is quick to point out that the restaurant industry isn't known for being a first-adopter of the next anything, whether it's in technology or any other trend. He cited a few pivotal points in the history of the industry when entirely new service options were made possible because of a cool gadget, like the drink coaster that vibrated when your table was ready, or the beginnings of the Twitter-based food trucks, like Kogi's in LA.

"I keep thinking about the next wave in the guest experience that could graft the power of an app for online, customized ordering with a fun in-restaurant sit down experience. It could shift the whole dynamic of the food preparation and diner" (Cardwell, 2011).

Fast forward from today and it's only a matter of time before someone combines the visual appeal of the iPad with mobile technology and social media links for something that resonates with the P.F. Chang's brand in the future. Stay tuned.

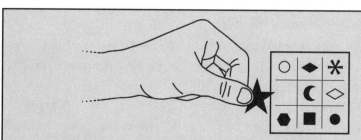

Go through Your Own Sort and Match Steps to Next:

- Don't wait for an emergency to develop a "next." The best options for a next come when your brand is strong.
- Theoretical ideas about a "next" are only as good as the ability of leadership to match the "pull of possibility" of the new with the "successful culture" of the now. Don't kill the goose that's laying your golden eggs as you build your next. Rick Federico said that as they grew, they always had to check to make sure the concepts were not "outpacing human capital."
- Learn from other industries. Financial services expertise taught Mike Welborn how to build scale into their concepts.
- Teach other industries and collaborators. The hospitality and restaurant industry embedded its core capabilities for driving employee motivation and rewards that took the concepts through periods of growth and change.
- Know yourself and others will know what you stand for. The exploration of a Japanese tavern wasn't a good fit for the P.F. Chang's brand. Strengthening the brand and taking it to global markets was a win/win. Stick to your mantras. Attractive distractions can come courting like sirens, luring you down a path you'll need to re-think.
- Be open to surprises. See an opportunity when it knocks, like P.F. Chang's partnering with Unilever to enter the frozen meal market.

SHARP HEALTHCARE+ HYATT HOTELS AND RESORTS

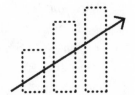

Staying Ahead by Improving the Invisible: How Two Organizations Discovered Untapped Opportunities in Guest-Customer-Patient Experience

Ideas are a dime a dozen. Our organizations could swim through the waters of trends for ages, observing with tremendous interest the infinite opportunities to innovate. Rarely do companies suffer from a lack of awareness about what could be, or a lack of great ideas for bringing a new level of relevancy to their brands and a new source of sustainable growth.

But, as leaders, how do we sift through those possibilities and commit to a direction for change that will have traction and differentiate us with customers? How do we discern, from a collection of evidence and hunches, the direction that won't redirect? How do we piece together the patterns of what might be and develop the focus of what will be? How do we pick from all of the choices?

The stories of two different organizations, Sharp HealthCare and Hyatt Hotels and Resorts, both committed to staying ahead of the trend curve, reveal the truth behind how we can really translate the art and science of ideas into sustainable growth strategies. Each of these examples is centered on one of the hardest aspects of organizational change: developing competitive advantage, brand distinction, and a new path to substantial financial results based on something invisible to the eye but mission critical to each organization. In the case of Sharp HealthCare, its invisible destination was to transform the overall experience for patients, families, healthcare workers, and community leaders. For Hyatt Hotels and Resorts, the invisible destination was to transform the overall hotel experience for guests, ultimately focusing on the health and wellness of people not in the comfort of their own home.

Sharp and Hyatt reinforce the notion of cross-industry insight as a way to change the competitive game. Both organizations read between the lines to piece together elements of the unexpected that led to dramatically different experiences for its patients (Sharp HealthCare) and its guests (Hyatt Hotels and Resorts), resulting in a deepening customer loyalty to their brands.

Sharp HealthCare Ignites a Wildfire: the Organization Takes on the Challenge of Creating the Best Healthcare System in the Universe

In 2001, Sharp HealthCare, a San Diego-based integrated healthcare system, accidentally set a wildfire. They gathered over 10,000 employees in a large room at the San Diego Convention Center and put forth a challenge: how could they move beyond their levels of service, technical capability, patient success, and medical excellence to significantly transform their organization? Meetings like this occur at many organizations, sometimes designed as a love fest to pump people up or as part of a regular rhythm of management communication, a one-stop-shop of messages to set the tone for the coming year.

But that particular meeting was different. Not only because Dan Gross, who started his career with Sharp in 1979 as a nurse and by 2001 held the title of chief executive offer for the Sharp Metropolitan Medical Campus, was a key leader and advocate of the day's events, but also because a palpable effect of the message ran through the day's presentations and discussions.

On that day, Dan, trained as an RN with the diagnostic talents of triage in his core, sensed that the mood in the room was different than anything he had witnessed in prior years. There was a readiness among the employees to bring the organization to an entirely new level. People in the organization were ready for change. A challenge was extended to the entire roomful of healthcare professionals: How could Sharp HealthCare integrate the new realities and trends facing healthcare into a bold new vision? How could the entire group of employees engage with each other and the community to understand the perspectives of their healthcare peers, with the families who would be served and the leaders who could support and inform the design of a new approach? Which metrics would be important to consider? What would it take to achieve a bold new vision?

Dan and his executive colleagues had led numerous employee meetings in the past and had always been impressed with the level of engagement of the assembled group. As he observed the mood in the convention center, Dan was struck by the staff's unprecedented high level of zealous support for that day's message. The employees were ready to jump on board in a big way. The first sign was the overwhelming outpouring of backing that came forward in response to the invitation to be part of task forces that would tackle the challenge of creating a new vision for the organization. One thousand people stepped up on day one to form one hundred cross-functional teams that would look at virtually every aspect of the organization's purpose and the possi-bilities for the future. The initial groundswell of support developed into initiatives over the next few years that would lead to Sharp being named a "Best Place to Work" and to the company winning the Baldrige Award in 2007, claiming dozens of awards for clinical excellence, and attracting extraordinarily strong support for a capital campaign.

The path that Sharp traveled was informed by field trips and case studies to understand the inner workings of companies including Ritz-Carlton, Disney, and G.E. Ritz Carlton mentored the Sharp team on specific skills that would help healthcare workers establish the type of eye contact and personal engagement that fostered a true connection with the patients and visitors. At Disney, they learned how to separate behind-the-scenes logistics from the public and separate patient and visitor-facing interactions. And G.E. demonstrated the importance of internal processes and the need to rethink metrics, rather than focus solely on clinical outcomes.

I journeyed through the Stephen Birch Healthcare Center at Sharp Memorial Hospital, the facility that opened in 2009 as the culmination of more than a decade's worth of insights about the new fundamentals for a healthcare experience for patients, family members, healthcare workers, suppliers, and virtually everyone who would interact with the Sharp community.

It was what I didn't see, hear, or smell that struck me first: no hustle of hospital beds being moved through the hallways, no antiseptic smells, no intercom announcements paging doctors, no waiting rooms filled with anxious people holding clipboards, no lines of IV poles in the corridors interrupting the sense of order. And, the way-finding signage was a dramatic departure from the typical color-coded and confusing arrows with indecipherable, technical terms.

The people that appeared on the scene reminded me of something altogether different than what I was used to in a typical hospital setting, that feeling of being in a busy New York deli, holding a number, waiting for my turn in an endless line. Instead, at every step of the onsite tour, the hospital staff knew I was a visitor, and seemed trained to tune into the fact that I might be lost and need help.

On the door of each patient room, instead of the typical ransom-note of assorted loose-leaf sheets of paper and clipboard charts, was an elegantly simple plastic sign with sliders that reminded me of the red and green flag system they use at the all-you-can-eat restaurants to signal when you want a waiter to bring more beef to the table. The Sharp flag system was designed to indicate to visitors when a mask was needed to enter that particular patient's room, and if the visitor needed to check with a nurse before entering. By the bed was a picture frame that had been pre-loaded with digital pictures of the patient's circle of friends and family. And there were simple but brilliant touches like sliding doors that separated the patient's sleeping area from the bathroom—something that allowed for easier maneuvering with walkers and wheelchairs and lowered the incidence of injury—and a sleeper sofa right in the patient room—for the family member or loved one who needed rest.

I knew that I wasn't in a typical hospital.

Then, the true magic revealed itself. I started to sense that the healthcare facility had been optimized for something I wasn't used to—healing—including aJapanese Jade Garden, artwork lining the walls of the hallways, and patient rooms that were not shared. In the visitor restrooms, inspirational quotes were stenciled onto the walls. Near the information desk you could hear classical music, and a piano was nearby for the impromptu player or the prearranged informal performance. The lights on the ceilings were offset from the center of the hallway. (Dan explained that when a patient is being transferred for tests on a rolling bed and he or she looks up at the ceiling, the offset configuration eliminates the harsh glare of bright lights that shine directly into the patient's eyes.)

Looking down from a patient's room onto the green garden below, you can see plantings that represent musical notes—the opening phrase of Beethoven's *Ode to Joy*.

Touches that could have been dismissed as luxuries were clearly viewed as a core component of Sharp's healing environment; its processes were designed to encourage a peaceful return to health. Every patient has a private room with its own bathroom and a comfortable place for a family member to sleep. That's because Sharp has evidence that patients who don't share rooms and experience fewer interruptions during the night sleep more soundly, request fewer pain medications, and heal more quickly.

"I have found that if you love life, life will love you back." —Arthur Rubenstein; inspirational quote, chosen by a Sharp experience team member stenciled onto the wall of the public restroom, Sharp Memorial Hospital.

And the spirit of continuous improvement clearly resonates with the staff on a daily basis. On a tour with Gross, we ran into an ER manager who was beaming about the most recent accomplishment ofher team. They had figured out a way to shave an extra twenty minutes off the waiting time between triage and procedures, something she said that had been at the top of their patient satisfaction scale. She understood that clinical results were not just a result of what goes on when the patient enters the procedure room; the waiting room experience can have just as significant an impact on the overall outcome.

With "healing" as the driving force, Sharp HealthCare had intentionally transformed the invisible—the hard-to-nail down but critically important quality that changes everything—to the actual experience of interacting with the clinicians—the tests, the metrics, and the health standards that add up to an improved sense of well-being and a faster recovery.

Measurable Results: From the Waiting Room to the Bottom Line

Patient Health and Wellbeing:
- Received county stroke certification at all four acute care hospitals and national stroke certification at two of the four (all four hospitals earned awards from the American Heart Association)
- Achieved a 73 percent reduction in catheter associated blood stream infections to a rate of 0.5 percent of central line catheter days
- Decreased hypoglycemia rates in hospitalized diabetic patients by over 20 percent to a low of 3.5 percent

Healthcare Standards:
- Overall patient perception of quality increased 80 percent
- Increased from 28 percent to over 90 percent "perfect care" across all national healthcare standards
- Sharp Memorial Hospital patient satisfaction increased from the thirteenth to the ninety-ninth percentile over the past ten years
- Sharp's Sharp Rees-Stealy named California's top-performing physician group by Anthem Blue Cross and PacifiCare, two of the state's largest health plans
- Sharp's Sharp Rees-Stealy and Sharp Community Medical Group awarded elite status, the highest possible designation for quality of care by the California Association of Physicians Group

> Dan Gross speaks on the role of leadership in change: "If Sharp Memorial were viewed as 'Dan's hospital' or if the changes we've gone through were imposed on the organization, we would have had success but not this level of raving belief."

According to Gross, you can't dictate things like the authentic desire of medical staff to donate the cost of triage rooms to a new emergency facility, a phenomenon that is memorialized with plaques on the walls of virtually every procedure room in the emergency facility. You can't issue memos outlining recommendations for improvement that would lead to floor nurses and other staff members collaborating on inspirational quotes in the public restrooms. That level of engagement and sense of ownership comes from an authentic desire on the part of employees to see Sharp's impact on patients and families from a new point of view and to take everything about their job to a new level.

Gross says that they began years ago by asking the entire employee group an audacious question, "How can we create the best healthcare system in the universe—from the perspective of patients, healthcare workers, and the community?" It was the boldness of that challenge that Dan believes inspired new thinking, and it was the sense that the organization wouldn't settle for less than that, that has kept the company's standards rising for the more than 11 years that followed. In Gross's view, "Phase One initiatives can start with process diagrams and flow charts, but the longer-lasting impact is to actually change the entire culture. That has to come from an internal belief on the part of every employee that what we're doing matters and that their contribution will have an impact."

Hyatt Addresses the Ultimate in the Invisible in Order to Exceed Guest Expectations: Air as the Next Frontier in Hotel Guest Room Experience

Hyatt's roots can be traced to a contract on a napkin when Jay Pritzker bought a Los Angeles airport hotel from a Dutch businessman in 1957.

A decade later, it distinguished itself through a new aesthetic with atriums, designed by architect John Portman, the first of their kind complete with indoor greenery. In the years since then, the hotel industry would shift in significant ways and become crowded with solid contenders for the top spot in guest preferences. The tastes of business guests, families, staycationers, and destination seekers would evolve significantly over the years and the choices of where to stay would blend into sameness.

Multiple high spots of hotel differentiation would come in the form of the Hyatt's Sunday brunches (a lure for locals) and their introduction of technology-enhanced guest services like self-serve kiosks and "line busters," hand-held checkout devices that could shorten the wait at the front desk. But, what could the Hyatt do in 2010 that would reestablish the impact of its brand in a similar fashion to Westin's 1999 hallmark introduction of the Westin Heavenly Bed, a concept that changed the conversation about the entire hotel experience to focus on the bed itself as a differentiator? How could Hyatt "own" an equally significant aspect of the guest experience?

On the heels of a recession, faced by an over-crowded competitive landscape, it would be hard to imagine what the Hyatt brand could do to land on its next direction. One thing it did know was that it wasn't worth the investment if the company tried to establish distinction with something obvious that could be easily copied. Hyatt's team had to dig into the drivers behind their guest satisfaction scores: what could take Hyatt guests from a level of satisfaction where Hyatt was as good as or better than its peers, to a level of desire, where Hyatt would stand out from the pack?

Hyatt's team sifted through mountains of guest data and satisfaction scores, studied general consumer trends, and looked at other industries from Apple to Starbucks to uncover something that could spark a substantial change, an approach that would offer the guests something that they couldn't imagine themselves, but that would truly distinguish Hyatt as a brand that is in touch with evolving guest priorities.

On what turned out to be a historic day for the Hyatt organization, a group of corporate leaders sat together to review ideas for strategic growth. They were in search of potential "delighters" that would wake the guest up to something special about Hyatt above all other hotels. In the room were three very seasoned hotel managers, Tom Smith, who at the time was working in the corporate offices in Chicago, Matthew Adams from New York, and David Nadelman from San Francisco.

Did Your Room Smell Clean and Fresh?

Every industry has a proxy for what to keep an eye on as the canary in the coal mine to alert them that something's awry. A large steakhouse I once worked with was in love with its certified Angus beef, when all the while the metric that really mattered had nothing to do with its prized cattle; one insightful team member fretted over his restaurant's declining scores on "clean restrooms" and knew that low ratings in that category heralded an era of a major slowdown in guest visits. Matthew Adams, vice president and managing director, had decades of experience to inform his understanding that of all the hotel customer satisfaction scores were reported to the Hyatt, one measure had the potential to be a game-changer: "Did your room smell clean and fresh?" Adams had observed over the years that overall positive guest scores in the entire cleanliness arena could be linked to whether or not a guest believed that his or her room smelled "clean and fresh."

Over the years, Hyatt had tried a number of approaches for hitting high marks in the clean-and-fresh arena from special scents delivered directly through the air conditioning units to new types of cleaning products. But, on the day that Pure Solutions presented to the Hyatt team, the pump was primed for Adams when the cleanliness-focused company started its pitch about a potentially breakthrough way to achieve cleaner, fresher guest room air than ever before. They had him at "hello."

The Pure Solutions team framed their presentation with findings from a 2006 hotel industry study that revealed an emerging trend: the majority of respondents would actually prefer a hotel that had allergy-friendly rooms and they would even be willing to pay a premium for them. Then, they presented their concept for building hotel rooms that were designed for cleaner, fresher air than had ever been possible.

After the pitch was the post-pitch reflection, which included conversations with John Wallis chief marketing officer for Hyatt. He and
Adams went back and forth about the appeal of the next frontier in the
guest experience. Starwood had introduced the Heavenly Bed and
changed the game for room amenities. Adams finally said, "What if we
could promise the guests a cocoon-like experience of being in an amazingly refreshing room? What if air quality could truly differentiate our
brand? "So Wallace stayed in the Respire Room in a Hyatt in downtown
Chicago and experienced the difference in the quality of his sleep. He
came in the next morning and said, "I think you're on to something—
the cleaner air experience has the potential to raise the bar significantly
on how our guests feel. Let's go for it." Shortly thereafter, the Hyatt team
committed to testing the concept in San Francisco and New York to see
if it had legs.

David Nadelman, general manager of the Grand Hyatt in San
Francisco, describes the evolution of the Hyatt Respire Room. "It takes
an entrepreneurial organization like the Hyatt to provide the Petri dish
for innovation, where teams are encouraged to be on the lookout for
concepts that can be game changing. The Respire Room concept was
actually born out of a long-standing understanding that the core of our
brand depends on our collective ability to continually up the ante in our
guest experience with something that makes a tangible, meaningful difference to their likelihood to choose us over the competition and to
enhance the strength of our relationship with the guests. But, to be
honest, we weren't specifically looking for cleaner, fresher air as the
new frontier until the serendipity of meeting the Pure Solutions team
appeared in front of us."

Nadelman and his colleagues receive dozens of ideas, proposals, and
products every week, and most of those pitches and concepts are never
acted on. But, every once in awhile, as was the case with the Respire
Room, an idea appears that represents something that could establish
the Hyatt as a company that is a step ahead of the competition when it
comes to what resonates with guests. Nadelman reflects on what was
different about the response to the Pure Solutions pitch that caught
almost everyone's attention. "We kept wondering if the air purification

system could set a new standard for hotel rooms, like comfortable beds had done a few years before."

At the very least, incorporating the Pure Solutions proposal might strengthen the Hyatt's guest satisfaction scores. But what was more compelling was a fascinating fact that emerged during the presentation: it turns out that there is an enormous number of people who suffer from allergies, asthma, or just sensitivities to environmental irritants—some estimates suggest it's a quarter of the population. And, based on the data that Pure Solutions presented, a large number of people don't know they have this problem, but they do know that very often when they stay at a hotel, they're not able to get a good night's sleep.

What if Hyatt could actually provide people with a really improved night's sleep?

The team realized that would represent a good opportunity for Hyatt to build its competitive advantage. And by being the first to do it—and doing it exclusively—they'd also have time to perfect the concept, so that even should someone else in the industry look to copy it, Hyatt would still have the advantage.

If all of the factors fell into place perfectly, Hyattcould achieve the trifecta: delighting the guests with the best night's sleep they've ever had, demonstrating that Hyatt wants to stay a step ahead in designing a great guest experience, and driving people to choose Hyatt over and over. If Hyatt could lead people to a better overnight experience than guest seven thought was possible, that would be a big wow.

Tom Smith, vice president of rooms for North America, described the six-step process that they designed, one that became the foundation of the Respire Room. According to Smith, "The air handler is completely disinfected and cleaned. Then, there's a tea tree oil cartridge put in the air handler in each guest room. All of the soft surfaces are completely extracted and shampooed. The room is completely detailed from a cleaning standpoint. Every surface is wiped down. Then, there is a product that is patented by Pure called Pure Shield that goes on all of the surfaces: tile, grout, carpet, soft surfaces. It preventsthe allergens from being able to land on any surface. It has been described to me as a cactus-like surface. When these allergens or pollutants land on it, it punctures them. It kills them on contact. One of the benefits is that a normal cleaning is sufficient to keep the Respire rooms clean for a stretch of six months before Pure comes back in and retreats the rooms. Additionally, there is an air purifier that circulates the air between 12 and 20 minutes every hour. Three to five times an hour, the air is completely circulated in what is classified as a high efficiency particulate air (HEPA) filter medical device. It's a four-step cleansing process for the air in getting rid of pollutants. We also encase the mattress and the pillows in hypoallergenic encasements to keep them sealed off. It is finished off with a certificate that's in the room that classifies it as a hypoallergenic room."

Respire Rooms Roll through New York

In the fall of 2010, a see-through rolling hotel room designed with all of the Respire Room touches, and complete with a live guest, steered its way through the streets of New York City. Branded as "Respire by Hyatt: Hypoallergenic Rooms," the concept turned heads and created the reaction reminiscent of V–8 ads ("I could've had a V–8!"). Smith explains how it's about more than just a mattress: "Yes, sleeping comfortably in a hotel room—it's not just the bed that would create an over-the-top experience—it's the air, the sound, the smell, and the entire ambiance."

How the New Standard in Guest Comfort Achieved the Wow

The Hyatt team launched market tests and introduced the Respire Room, which led to very high marks from guests. The Respire Room now represents 2 percent to 5 percent of Hyatt's room inventory. Guest satisfaction surveys reveal that for a number of guests, the Respire Room experience is the first time they've gotten a good night's sleep at a hotel. And, even though many of the guests never thought that cleaner, fresher air was part of the equation, after guests experience the Respire Room, they realize how important clean air is to achieving a great night's sleep.

The Respire Room also did exactly what the team had hoped for in helping Hyatt achieve a competitive advantage. David Nadelman saw early signs that they were onto something big, "We were the first hotel in San Francisco to have this kind of room and right now the only hotel company to have this as a brand standard. People have started to request and pay for the upgrade and it's driving repeat business."

The Next Next

Tom Smith sees even more impact ahead for the Respire brand. He is beginning to envision potential synergies between that concept and some other trends in which Hyatt has led the way, like "Hyatt at Home," "StayFit," and "Healthy Living." The company's Hyatt at Home product line uses an online catalogue of offerings that extends its brand and

Matt Adams Envisions a Whole New Travel Journey

Ask Matt Adams what he's got his eye on for the future: apps and sites like Coolhunting.com for new trends and gadgets that are hitting the marketplace. Trends like locavore food culture inspired Adamsto install a 24-hour food market that is billed as an "instant gratification food and beverage venue" and which could replace room service as a top choice for guests. He also learned about corn-flavored ice cream, inspiring the chef to take a new direction with what he served Hyatt's guests. In addition, he wonders how technology like radio frequency identification (RFID) could help track baggage and how text messages could be used to create a much higher quality "travel journey" for guests. Imagine landing in New York City and being greeted by a customized message, "Ms. Kates, thought you'd want to know we're waiting for you. Your room number is xyz and your gold card has been preprogrammed to open the door." And, upon arrival, you'd be met with displays of the local culture: clothing designers, artwork, chefs, and urban leaders. So that you knew you were in New York and you felt that you were having a customized, local experience of the city.

"That kind of vision challenges you to look outside your own industry for inspiration and partners," Adams added. "That's how we'll create a truly distinctive 'next.'"

hotel experience beyond a guest's stay. It offers luggage, bedding, wine, and even furniture. Matt Adams explains that there has been a shift in perception about hotels over the past decade or so. "It used to be that hotels set their décor based on trends in home furnishings. Now, the opposite is true. People want to purchase the bed or home accessory that reminds them of the scale and style of the hotels where they stay."

Tom Smith is beginning to observe an emerging pattern that could develop into new corporate DNA for Hyatt. The StayFit line of fitness and wellness products combined with programs like the Healthy Living program could merge with the Respire Room insights in the future to offer a totally unique sense of overall wellness for guests. The possibilities in that arena led to the next set of opportunities: "What would happen if the Hyatt could have an impact on the entire travel experience from healthy eating to a great night's sleep, to keeping fit to working more productively, and extending that experience beyond the time you're at the hotel?"

Think of how many aspects of our lives that could ultimately be Hyatt-ized.

Go through Your Own Branding Steps to Next

1. Brand differentiation that sticks comes from a deep dive into data, trends, and experiences, combined with the art of intuition. Hyatt studied guest satisfaction scores and the success of its competition. The Sharp HealthCare team incorporated the perspectives of patients, their families, healthcare professionals, and the entire employee base into their new vision. Each uncovered opportunities to redesign the entire experience associated with their brands that blended the science of analysis with their instincts about untapped potential.
2. Test your concepts to take the risk out of the unknown. The discipline of Malcolm Baldrige-organized processes accelerated Sharp's ability to land on approaches that would make a significant impact on specific aspects of the patient experience (i.e., waiting time in the emergency room). Hyatt's expertise in

market testing every aspect of the Pure Promise to make sure that allergy-free science translated to an improved guest experience is what made the difference.

3. Model your standards for success after companies outside your own industry. Sharp HealthCare used Ritz Carlton as a model for the quality of the hospitality experience and Disney for behind-the-scenes logistics that would never trample on the peaceful atmosphere patients and their families needed to feel. The company created new metrics—different from those of other hospitals—to start to describe and measure dimensions of the healthcare experience beyond traditional ones like "length of stay," "mortality/morbidity," and "efficacy."

4. Keep your eyes open for opportunities to raise your bar and insert an element of surprise. Westin rose to the top position in the J. D. Power rankings in room satisfaction between 1999 and 2004 after it introduced the Heavenly Bed. Hyatt is trying to raise the bar again in a new dimension of the guest experience and it is on a continuous search for the new next.

5. Master "followership." Leadership of significant change initiatives can begin along a conventional path: uncovering information, inspiring the organization to rally, and empowering people to develop processes that will make the vision come alive. But ultimately, to quote Dan Gross from Sharp HealthCare, a leader needs to practice a fair dose of "followership" and listen very perceptively to everyone within the organization and tap into the power of the collective voice.

GENERAL ELECTRIC ECOMAGINATION / GENERAL MOTORS ONSTAR / EMC CORPORATION / KORN/FERRY INTERNATIONAL

The Art of Seeing What's Possible: Big Stakes, New Concepts, and the Opportunity for Transformation

Sometimes business crossroads come when we're least expecting them. Theart of seeing what's possible traces four companies—G.E., GM, EMC Corporation, and Korn/Ferry International—that were already on course and in motion of practicing that almost unbeatable skill. For different reasons, each came to the conclusion that its tomorrow needed to look different than its today. G.E., for instance, picked up on subtle signs that traditional manufacturing might be due for a shift, based on early indications—scarcer natural resources and a different ethos on the environment—that a new era was coming. It needed to weigh in on whether sustainability was merely a fad or actually a new reality. G.E. was looking to figure out how its business could thrive if, in fact, the environment began to play a more dominant role in the commercial equation.

Hughes Electronics and EDS (Electronic Data Systems) were part of the GM family in the mid-1990s, together creating cross-talk about evolving public sentiment around connectivity and security inside vehicles. The discussions were rich with what-if scenarios. How could each of their core strengths come together, and enable them to surmount a coming tide in automobile capabilities that went beyond engines and options? The GM team was faced with a two-part challenge: creating technology slightly ahead of its time and anticipating public reception to a revolutionary concept of car-driver communication and feedback.

EMC Corporation was founded on a platform of innovation. But after an initial wave of success, it experienced a reversal of fortunes in the early 2000s as the market shifted and one of its core areas of expertise—data storage—became a commodity. EMC's team had to discover what their next big move might be by stepping outside the realm of the day-to-day. It envisioned the future as a world that separated data storage from the computer that created it, leading the way for what would become the next significant change in the industry—the "cloud." But, the company would face challenging decisions to clear the way toward market leadership.

And finally, Korn/Ferry International provides a sobering example of what can result from an over-zealous response to a trend. In the heyday of the dot-com boom, its leaders saw a gloomy future ahead for their executive search firm. Korn/Ferry's people watched as its sacred currency—the executive résumé—started to leak out onto generic Websites. They feared their company was heading down a path toward extinction. In their overreaction and eventual recovery lies a universal truth, and Korn/Ferry's story offers a powerful and refreshing vantage point on the dos and don'ts of trend spotting alongside innovation.

Together, these four companies illustrate the art of seeing what's possible, in startling different, yet similar, ways. Picture a series of debates on climate change. Not academic debates with a trophy at the end of the battle of wills, but a highly charged corporate environment with a roll-your-sleeves-up level of passion and engagement. At the end of this kind of debate isn't a consolation prize, but a solid stake in the ground and a road that will lead to a multibillion-dollar initial uptick in sustainable revenues.

GENERAL ELECTRIC ECOMAGINATION

Train Your Gut: Immersion and Integration

Mark Vachon, vice president for G.E.'s ecomagination division, was in a completely different role when Jeffrey

Immelt staged exactly that type of debate in 2004. The competition began when Immelt noticed several divisions working on independent projects in energy efficiency while other large corporations were starting to incorporate environmental issues into their strategies. He charged two teams of experts with taking on the huge topic of climate change, and performed in-depth due diligence on every aspect of the equation, eventually presenting the findings to all 35 senior leaders.

The response to the notion of incorporating a "soft" issue like environmentalism into G.E.'s traditionally hard-edged culture was far from exuberant; in fact, the entire executive team voted the idea off the drawing board. Vachon recalls, "I remember the look on my face: it telegraphed a very healthy dose of skepticism about the very premise. I couldn't help but wonder why a company like ours, with a tradition of solid process design and expertise in things like building and running coal-fired turbines, would be investing so much thought in what seemed at the time like an offbeat notion, that the future held tremendous benefit if we could deliver a game-changing level of environmental performance."

But Vachon and his peers listened and learned, and over the course of the extended series of leadership meetings and months of continued communication that followed, they began to see that G.E.'s shift in course was destined to become a new imperative. The debate taught them that the elements that drove G.E.'s strategic equation for the past 130 years were in a state of change. Resources were scarcer, even as emerging economies were growing at accelerated rates. Not only was science supporting the position of a fundamentally new outlook, but public sentiment on a global basis was starting to weigh in and it looked as if the tide was shifting irreversibly toward sustainability.

The G.E. team was challenged to figure out how they could reduce G.E.'s own greenhouse gas footprint by 20 percent, significantly improve its own energy efficiency, substantially reduce its own water usage, while at the same time, move aggressively into markets in which it could lead the way in clean energy. But, once the debates were over, and Immelt's persistence on the topic finally resonated with the G.E. executive team, they committed to the task. And once they did, the question no longer remained whether or not G.E. would invest in this new, uncharted path, spend billions of dollars, and master the art and science of leadership in a major new business. They were left only with the "how."

"In an organization the size of G.E., there were outrageous goals and we had no idea how we were going to achieve them. But leadership's role had been to communicate the 'why' and then to establish seemingly undoable goals and get the organization to step forward and accelerate success. I had no idea that just a few years later, I would be asked to lead the charge in what evolved into G.E. ecomagination," says Vachon. Ecomagination became a companywide business strategy that centered on the ambitious objective of figuring out the economics of sustainability, and providing novel, market-leading technologies in that arena on a global basis.

"In the end, ecomagination was only partly about saying 'We believe in the signs,' which wouldn't have led to a business focus. The true significance was in declaring our belief in the markets we could create—trillion-dollar markets in wind, solar, and water. Jeff Immelt demanded that from that point onward, each business would have to develop products, services, and solutions that were not only responsive to the market in terms of economics, but responsive in terms of environmental performance. It was bold. A lot of us were surprised and confused. But now, looking back at the imperatives we all integrated into our thinking, it looks brilliant."

The organization has, in fact, figured it out. Ecomagination is now at $85 billion in revenue and it has initiated lines of business all over the world that lead its markets in environmental sustainability and that are highly profitable. Innovations like a steel factory in Macedonia powered by the country's first gascogeneration plant, a Chinese poultry waste biogasenergy plant, and a water system in Green Bay, Wisconsin, that brings new efficiencies to water power have set the stage for the coming decades where G.E.'s ecomagination can continue to mine ideas for sustainability in energy.

Very few of us will ever be in a position to re-steer a behemoth, 130-year old ship like G.E. toward an uncharted, new direction, but all of us can relate to that moment when we sense that our company is at a crossroads and the stakes are high to get it right, especially when the new focus requires a leap from where our current business is to where we envision the potential for the future.

The process for knowing where the opportunity lies can start off with tremendous feelings of discomfort, as was the case with the G.E. team. Before Vachon was tapped to head up ecomagination, he'd spent time as the CFO of NBC, one of G.E.'s non-manufacturing entities.

"Believe it or not, my level of confidence with leading in an area where I'm not 100 percent comfortable came when I had to apply the rigors of finance to a world where the spreadsheets couldn't provide certainty for what a blockbuster comedy would be that season or what a great drama was. I surrounded myself with people who were seasoned in the fundamentals of the business of television hits and learned through immersion how to see what they were seeing about what was possible."

Steering the Dart Toward the Bull's Eye

The difference between a summary sketch of how conclusions are drawn and one that connects the dots of data to insight is like the difference between a football playbook and the quarterback's ability to think and move nimbly to score the touchdown. You can listen to comedy writers describe plots and characters all day long, but you can't know what makes them funny. Mark Vachon offers similarly intuitive insights on how to place a bet. "There is no spreadsheet that tells you what a funny comedy is or which great dramatic story will resonate with the public. You have to surround yourself with people who have experience and stay in close touch with trends in the market. And, in my case, I rely on my gut instinct, which has been trained by 29 years of experience ... All of the drills that you do in the early days of your career expose you to the fundamental physics of business, whether it's in compliance, controllership, regulation, and strategy. As you apply those fundamentals to bigger and bigger sandboxes, you realize that the elements combine in new ways every time, but your basic conditioning has served you well—your instincts respond to ambiguity with a better sense of what's right."

At large companies like G.E., leaders have the added advantage of integrating ideas from one division into another. As the ecomagination initiatives in the area of wind-powered energy generation hit a wall, G.E.'s management turned to the company's aviation group for answers. What do you do when the wind dies down? The aviation engineers had solved the problem of how to flex jet engine power up and down extremely quickly. G.E.'s F50 turbine became the perfect hybrid between aviation and wind-generated power, conceived by two parents that were brought together to solve a new challenge. "Business leaders today have to do a constant scan of the horizon to absorb enough information to stay ahead. At G.E., we're lucky to have a huge menu of resources at our disposal—an amazing richness of employees, advisors, researchers, and a broad range of projects that can combine to form new core capabilities. It's the transfer of thought amongst our businesses that pushes us out in front of the future much more powerfully than an a la carte approach would."

For the rest of us not running multi-billion dollar divisions within mega corporations, we can treat the world of commerce as our laboratory by arming ourselves with knowledge of the physics of our own companies and instilling scanning discipline into our leadership playbooks. Where else could tomorrow's solutions come from? We can all do what Vachon and his team did as they immersed themselves in the worlds of aviation, locomotive transportation, capital, and energy, and focus on the questions that the experts in each of those disciplines is thinking about. "I'm always asking myself what I should be thinking about to accelerate progress and scale up faster. In the pre-NBC stage, when I operated inside more familiar territory from year to year, I might have been more internally focused and even scrunched my face when I listened to what the global research center teams were working on, thinking, 'Why are we working on that?' Now, I realize that tomorrow's opportunities come from connecting pieces of a complex puzzle. It's my job as a leader to see what the researchers are seeing and fit the parts together to shape our direction."

GENERAL MOTORS ONSTAR

Finding Opportunities by Examining What's Not There

Nick Pudar, who has driven OnStar's strategy for years, echoes Vachon's view of how to learn to see what's not there and tap into what you envision as the big unmet need. Pudar believes that the process of discovering what's possible always starts with a novel question and a fresh perspective. His initial inspiration came from a story about one of the big soft drink companies, which called for a timeout in the middle of a strategy session about winning market share from its competitors. According to the story, the product manager turned to her peers and said, "We've spent three days talking about the 16 ounces of soft drinks that people are drinking every day. If a 150-pound person is taking in 60 ounces of fluids a day, that means there are 44 ounces of liquid he's drinking that is not a soft drink. What are we going to do about that extra 44 ounces?" And, so began the era of bottled water, tea, energy drinks, and a new generation of liquids to fill up that drinkable space.

Pudar points to such a precise line of questioning when he describes the inception of OnStar. "Early in the days of telecommunications, the GM teams were talking about what could be done to leverage the evolving capabilities of mobile technology. We pored through data, talked with our colleagues at Hughes Electronics and EDS (Electronic Data Systems), and landed on a big idea. We envisioned the transformation of the car into a space where people could feel completely connected and secure." The idea was compelling—to transform the experience of driving into an environment where drivers didn't have to worry about being in an accident with no way to get help or getting lost on unfamiliar roads without assistance.

At the time, the vision of providing an unsurpassed level of connectivity wasn't achievable technologically. The cellular infrastructure wasn't robust enough to support the vision, and it was cost prohibitive to build it out. But the early team was firmly convinced that the concept would be game-changing for drivers. They were so committed, in fact, that they cobbled together prototypes to test the idea in the market as quickly as they could with existing tools. OnStar's early development group focused on building what would become a few generations of iterative solutions, starting with the infrastructure that was available at the time, and refinements in sophistication evolving at every step. Pudar looks back to the crossroads of the idea's inception and thinks about how their conviction evolved into the eventual solution. The scales tipped in favor of OnStar's market launch once the group cracked through an assumption that they hadn't realized was a limiting belief to the entire rollout.

"The fascinating thing about assumptions is that a lot of times what most people characterize as a given simply isn't, but it takes a very different vantage point to realize it," Pudar explains."I'm trained as an engineer, and I pride myself in understanding the basic realities of what makes things work. To get OnStar to market, we finally had a eureka moment. We realized that we were analyzing the timelines like car people, with the rules of manufacturing as our givens. There was a historic day when we woke up and realized we had to think like consumer electronics people instead. We knew we had to have the OnStar product move from a dealer-installed system to one that was installed in the factory, in order to gain all the scale economy benefits. However, the typical automotive development cadence at the time meant that we could start with only one vehicle, and that it would be five years out. That was clearly not in the necessary consumer electronics time horizon. The OnStar team, in conjunction with some creative vehicle engineers, started from scratch, and figured out how to workout the development timing mismatch. In just a few short months, GM was installing OnStar on half of its vehicle portfolio. And that penetration level increased to nearly 100 percent in just a few years."

Today, 6 million subscribers later, Pudar and his team have proven that their perceptions about the potential to unlock a new market need had been right on target. OnStar is the leader in connected drivers providing each other seamless experiences while keeping both hands on the wheel, a concept that is truly forward thinking. But, Pudar also learned that the definition of connectivity continues to change and influence where the business is going, even today. Having the ability to see the cues, possibilities, and potential around him opens more doors than looking forward does.

EMC CORPORATION
Jumping Ahead of the Cloud

You enter the offices of EMC Corporation in Massachusetts and immediately sense that some big ideas underlie the activity and the whir in the hub of a company that employs more than 50,000 people worldwide. William (Bill) Teuber, vice chairman, reveals the spark behind the ingenuity as he traces the company's evolution from a three-man start-up in 1979 to its position today as one of the world's most admired technology companies. Teuber can talk about the same concept that inspired Mark Vachon of G.E. and Nick Pudar of OnStar—how EMC discovered its next by spotting signs that the business landscape was shifting, and then pioneering initiatives to accelerate the company's ability to lead in an entirely new playing field. Teuber's story begins with a rise and fall of fortunes.

From its early days, EMC distinguished itself through innovation, with its first significant demonstration of creativity in the conceptualization of an elegant computer storage product. Teuber describes EMC's innovation DNA: "Those were the days of single large expansive disks, or SLEDs, and there was no independent market for information storage devices. In 1990, after three years of research and development, EMC introduced Symmetrix, a storage platform that represented truly ground-breaking thinking about the entire category. The year before, EMC had been a small $100+ million company focused on providing memory boards for computer systems from the likes of IBM, DEC, HP, and others. Then, EMC leveraged its expertise in the computer memory business to develop hybrid storage systems that utilized both memory and disks. We basically created a category of storage that was separate and distinct from which server you were using."

That move established EMC as a maverick, and within months they were competing and winning against IBM and Honeywell. Teuber was the independent auditor who signed off on the 1990 financial statements for the company. "As I was doing the review of the financials, I said to the second partner, 'This is a billion dollar company waiting to happen. You can just see it.'" EMC closed the year at $170 million and

Teuber's instincts proved right. Four years later, EMC was a billion-dollar company that had earned the reputation as category leader.

In 1995, Teuber came on board as a full-time member of the team, largely attracted to the company's roots in innovation, and was part of the group that helped to lead EMC to become the fastest-growing stock on the New York Stock Exchange during the decade from 1990 to 2000.

Then, in 2000, the dot-com bubble burst and the crash hit. EMC was forced to cut 24,500 people to 17,000 and reduce operating costs by a third. It did what it could to weather the challenges of that era by focusing on survival. What would a company that was founded on innovation do when the forces were against it?

EMC Turns the Tide from Downward Spiral to Industry Standard in Less Than Five Years

"EMC had a harsh wake up call in 2001 and 2002. We had to get ourselves right-sized; we had to weather revenues falling from almost $9 billion to $5.4 billion, which was about our break-even point. We had to face some harsh realities, but our strength came from understanding our company's core DNA." Teuber shared the insights he and his team learned from that retrenching period, including the importance of managing its cash, and the power of research and development for a company like EMC, with a differentiation that came from product leadership and seeing ahead of the direction of the market. Even during a downturn, EMC's leaders knew that cutting off their company's wellspring of innovation was not an option. "We recognized that the specific product solutions that got EMC to where we were weren't going to be sufficient to carry us toward the next phase of growth, and we knew that we had to keep investing in research and development, no matter what. We were unwilling to sacrifice the future of the company by cutting spending on innovation and research."

To change the tide on opportunities, EMC maintained a conservative percentage of money spent on R&D throughout the reorganization period, preserving cash through stringent financial management, so that it could establish a foothold in new areas of expertise when the timing was right. Over the next eight years (from 2003 to 2010), EMC would invest $10.5 billion in R&D and another $14 billion in acquisitions, to acquire smaller companies with complementary technology in order to build out EMC's product portfolio and extend its market reach. In 2003, EMC bought Documentum, a content management company, and the purchase was a first step to get ahead of what they saw as an emerging market outside its core offering in storage. EMC recognized the majority of the projected growth of data was going to be in the area of unstructured data—not the type of data that sits in rows and columns in a database, but in files of digitized video, images, and documents that could be stored and accessed in new ways. "Our team did studies. We were vigilant, watching the tremendous rate of growth of data. Unfortunately, we're in an industry in which the unit price goes down every quarter because of continuous advances in technology, so data growth alone would never drive profitability. We had to uncover a new direction."

The EMC team dived in to piece together the components that drove their company's market at the time. They talked with customers and learned of an appetite for more robust capabilities in software. The company's R&D group learned more and more about the challenges of mining unstructured data. But the management team sensed there was an emerging market need that would dramatically cause a shift within EMC's industry that couldn't be seen simply by adding up their existing spreadsheets and data analytics.

"Joe Tucci, the CEO, really understands where the trends in the industry are going, and he has been spot on in sensing trends on more than one occasion; he has a feel for the market and you can never substitute data alone for a feel for the market. You have to have both the data that lays out the situation and the gut instinct to piece together the information to drive toward an actual insight," says Teuber, who outlined the perfect storm conditions that led to EMC's purchase of VMware. That 2004 acquisition had some people in the industry scratching their heads. VMware virtualization software allowed a single server to operate as multiple "virtual machines" and multiple servers to operate a virtual pool of resources. Virtualization had the potential to grow into a foundation enabler of the next wave in information technology—cloud computing. "Most people didn't know what virtualization was at the time we acquired VMware. We were paying a lot of money for a company that had very little in terms of revenues. But it was Joe who saw the future by piecing together evidence from trends that were emerging in small ways throughout the industry. Joe had the foresight and the understanding of where the market was going. He envisioned the whole shift from data that was physically married to a server to a world in which data would be stored in what we now call the cloud."

The marriage was set in motion because of a second critical component to the vision—the realization that for virtualization to succeed in the marketplace, the major server vendors would have to embrace it. Like EMC's early, independent storage platforms, VMware's virtualization software would have to be seen as server-neutral. "We were in the perfect spot from VMware's point of view—they were looking for a liquidity event and didn't want to go to Hewlett-Packard or IBM because their virtualization software sat on everyone's servers; they wanted to be server-agnostic. We were an agnostic option for them. We made the decision to acquire VMware and to keep them independent of EMC, separate and distinct from the core storage business, because Joe understood that the real value was for VMware to become the standard of virtualization for the entire industry."

As of mid–2011, VMware's valuation was upward of $40 billion and climbing, and according to Teuber, it's been called the best corporate investment ever, with recent growth in the double digits as the industry begins to catch up with Tucci's revelation about the power of virtualization: the ability to run multiple server and storage devices as a pool of shared resources, and the underlying power behind the coming era of cloud computing.

Teuber admits that EMC might not have rebounded so strongly had they tried to grow only organically after the downturn in 2001. "It was one of those things where if you looked only within your own organization for the answers, you wouldn't have gone off and made a bid. If you only asked yourself, 'How can we grow incrementally?' you would never have taken that step. It was almost as if we had to swim through a pool of data, factor in what the customers were telling us about their emerging challenges, and look outside our own four walls to see who was thinking about the next wave—the cloud—and inject their capabilities into our own perspective. Then, follow the instincts of people like Joe who could help us all get ahead of the future."

KORN/FERRY INTERNATIONAL

A Word on Risk: The Downside of Innovation

Lest we go away thinking that every hunch is brilliant and the secret to success in business comes from trend-spotting and putting all of our eggs in the next new shiny basket, we know better when we meet Don Spetner, who, as executive vice president of corporate affairs for Korn/Ferry International, has a powerful story to tell. It's not so much a tale of doom and gloom, but a sobering moment to guide all of us through times when we're seduced into thinking that we need to dive in quickly or jerk our knees suddenly when we see a disruptive technology invade our competitive arena.

First, you have to hand it to Spetner that he's willing to share a story of early missteps, but now that his company has recalibrated and recaptured its number one market position, Spetner believes the elements of Korn/Ferry's experience can serve not only to remind the company how important it is to resist the siren's call sometimes, but also to illuminate the real risks involved as we are tempted to follow the rallying cry and allure of innovation and trends.

Korn/Ferry International is the world's largest executive search firms. Since 1969, its bread and butter has been its *retained search*, in which the company receives six-figure fees to find and place executives in top positions with corporations around the world. Thirty years after its inception, Korn/Ferry saw a big trend right at the company's doorstep: in 1999 the Internet had become a force with the potential to create significant new business opportunities for Korn/Ferry, and change the way it approached the talent market.

The fee structure for retained executive search firms is straightforward: they are paid a fee equal to 33 1/3 percent of a candidate's first year's annual compensation. Thus the fee for a job that pays $600,000 would be $200,000. While this is a lucrative market segment, it is also relatively small, representing less than five percent of professional jobs.

When the Internet exploded on to the scene in the late '90s, Korn/Ferry saw an opportunity to move into a much larger sector—the middle-manager market. With candidates willing to post their resumes online, it seemed a search firm could easily sort, identify, and present candidates at any level, and do so profitably.

Armed with the notion of taking a potentially disruptive technology and turning it to its advantage, Korn/Ferry made what it believed to be the right move. It raised $200 million in capital to fund its response to the Internet opportunity. The story to investors was based on a very credible belief that it needed to invest in getting ahead of the curve to not only avoid a credible threat to the existing business model, but to lead the way in to the future. Spetner looks back at their assumptions at the time and how they went south: "When the first Internet wave came, we unfortunately overreacted, and ultimately squandered significant dollars, trying to rush into the eddy of opportunities. We responded a bit too hastily, partly out of fear that we would go the way of the newspaper industry—losing our footing, possibly forever."

Following the dot-com bust, the team rethought their original assumptions and changed lanes. Korn/Ferry had to face the fact that its initial dot-com strategy was off course when things went south—clearly Korn/Ferry'sinvestments weren't playing out as planned, evidenced by such a dramatic downturn in its financial results. Just as EMC discovered components to its core business were becoming commoditized over time (in that case, data storage), Korn/Ferry looked more closely at its situation and concluded that, yes, résumés themselves were going to lose their luster as the differentiator in the world of executive search. However, the Korn/Ferry team determined that there was a silver lining in that cloud of bad weather.

"In 2002, we were extremely low on cash and in real business danger. We got a capital infusion from a private equity firm, brought in new management, and completely turned around our business. We also began to search for the answer to a very fundamental question, 'How does the executive search business really evolve, what should we do about it, and where does our strategy need to go?'" Spetner recalls. "We were forced to re-think the true impact of the dot-com dynamics on our business.

Instead of assuming our future would be like the world of newspapers, which had been the initial driver of our fears in the heyday of the dot-com era, we realized that our world might be more like the world of real estate, where for some reason agents had not disappeared from the picture."

First, just like EMC, Korn/Ferry right-sized the business to stabilize the company and looked for directions that would provide more value to customers. The team engaged in discussions about the basics of their business, even evaluating whether or not it should go private again and remove the growth requirements imposed by a public company structure. They talked to customers. They looked at brand extensions. "Basically clients came back and said, 'We assume you guys are experts on all kinds of things beyond just finding people. What about salaries, succession planning, performance development—everything associated with executive-level talent?' So we began to build a new strategy around being a talent solutions company rather than an executive search firm. That's what the last phase of our development has been about and I'm glad to say we have executed very well against that strategy."

When You See an Eddy of Trends, Wade, Don't Dive

Don Spetner shares a secret about disruptive innovation, a concept extremely well-suited for circumstances in which a company has an opportunity to create automobiles in a world of buggies. According to Spetner, "One of the big lessons we learned was that sometimes when an opportunity for disruptive innovation comes along, don't dive in head first—wade.

"We learned from our missteps during the dot-com era as we invested heavily in order to quickly seize a competitive advantage. We ultimately moved too precipitously. When we recovered from the first wave of our response, we were a bit more balanced in our view of the world: we looked cross-industry not at severely damaged sectors like newspapers, but at other industries like real estate as a proxy for a path we might follow.

"Real estate agents looked like perfect candidates for disintermediation. It seemed inevitable that real estate brokers could ultimately be replaced by electronic listing services on the Web. And yet, it did not appear that agent commissions were under attack. Smart real estate companies amped up their points of differentiation, expanded services like mortgage financing and inspections, and built a new value equation based on the specialized intimacy that an agent could provide throughout the home-buying process that could never be replaced by an electronic listing.

"We tried to apply those lessons to ourselves and began to extend our services to include assessment and the science of talent management and started down a path to become market leaders in those emerging areas."

Your Checklist for Grasping the Invisible ☑

1. Sort out where you are: take your pulse and check your liar's box. EMC watched as storage became a commodity and had the foresight to rethink their options. ☐☐

2. Combine elements to shape the future. G.E. read the shifts in perceptions about the environment as significant to its future and blended those insights with its manufacturing prowess to design a multitrillion-dollar commitment to a new corporate direction.

3. Learn the lessons of adaptation. Korn/Ferry and EMC experienced lows before their eventual rebounds. Create a culture that can flex and respond to shifts in the market.

4. Drive toward execution once the vision is in place. OnStar's team envisioned a future that didn't exist, in whichcars could be more than transportation alone. Jeffrey Immelt overcame internal resistance to change to shift the organization's focus to a new approach to manufacturing.

5. Don't overreact to trends. Korn/Ferry's experience taught it that it didn't need to leap into a new business model, despite a significant looming threat to its core business. When you see a trend, decide if you should dive in or wade first to change course.

Conclusion

Not every company will be as visionary or as lucky as EMC, or as bold as G.E. Many organizations will never have to think outside of incremental growth to survive. But, for companies that want to, or need to, rethink their current course, understanding the lessons of transformation can accelerate the speed of success.

We can all learn to see between the lines.

We can all ask fresh questions.

We can all tap into the combined genomic elements that make up our business' DNA.

We can all discover cross-industry trends in plain view and devise novel approaches to growth.

This final note on transformation brings us to the smaller scale moments when we feel the urge to make a new mark. Whether we're sitting at our desks thinking about how to penetrate a new market, or imagining how our napkin sketches could develop into a new business, we can follow in the footsteps of the individuals who broke out of conventional thinking and defined a new next, such as Danae Ringelmann, who quit her Wall Street job to create an entirely new system for funding independent, creative, cause, and entrepreneurial projects using crowd-sourcing—IndieGoGo.

"   "

"I knew there had to be a better way to get creative ideas and ventures off the ground than the traditional Hollywood models of Hollywood or Wall Street. I suspected that IndieGoGo's model would allow new voices to emerge in the independent world. But, I had no idea that the artists and entrepreneurs themselves, and to a certain extent, the works, would be so changed by the new dynamic, where fans and future customers would be engaged and rewarded as part of the process."

There's also Brad Inman, who set the world of publishing on a new course with books that integrate video and game technology to enhance the story—Vook.

"   "

"I had a notion that the devices we were buying like computers and iPads weren't keeping up with everyone's desire to experience content in a new way. When you read a book, you're bound to the structure of the page. But when you flip through your music collection or surf the net, you're able to pick and choose snippets and dig into each area of content based on your own preferences for navigation. With Vook, I was inspired to marry the two experiences—the narrative story line and the sense of exploration. So we created an entirely new way to experience content and formed a company around it."

Scott Wilson solved the chicken-and-egg problem of industrial design, raising nearly a million dollars through crowd sourcing to fund his Nano-based watch—LunaTik + TikTok.

" "

"So much great design never sees the light of day simply because the mechanism for bringing design to market has involved some hurdles that are hard to overcome. The LunaTik + TikTok experiment used Kickstarter to raise funds instead of venture capital, which sparked a whole new relationship between the product and the people who wanted to own it."

John Winsor shifted the dynamics of advertising to the crowd, creating a global community of creative talent who mobilize and recombine to form virtual marketing and promotional teams—Victors & Spoils.

" "

"The old days of advertising are heading the way of the dinosaur. Charging huge markups on advertising purchases is an idea that won't stand the test of time. Neither will the ad agency model where a small cadre of creative geniuses forms the basis of an ad shop's competitive advantage. The industry was ripe for an overhaul that unleashes the talents of the many, many creative artists, copy writers, project managers, and communications strategists on projects. That's why we formed Victors & Spoils, which is based on a reverse auction model. Now, a project is conceptualized by a client and then the community can collaborate, bid, and one-up each other's ideas, ultimately raising the bar on the caliber of the solutions."

Alistair Goodman went for the less-sexy-than-apps platform of text messages to provide place-specific, customized communications for customers who opt in to a service that tells them what they want to know about locally available products, right on their cell phones—Placecast.

" "

"As a start-up, we focused on two things that were hidden from plain view amidst the flurry of activity around the development of mobile apps. First, we realized that location itself was a powerful anchor to the whole mobile experience. The shift was subtle; like when you used to associate a land line with a house, except, now, you'd associate a mobile device with a person. Where that person is located contributes very powerfully to the new equation. Second is privacy. Trust and privacy are paramount when people opt in to a Web-based or mobile service. At Placecast we based our entire model on those two new givens and built a model that now has millions of people opting in."

And there are visionaries like Tim McEnery, and the place where our story began, the man with a hunch that he could build a new community of diners who felt so loyal to his suburban Napa experience they'd grow into loyal wine club members and travel club participants—Cooper's Hawk.

" "

"I remember when we were a single location winery restaurant and I was a sort of one-man band, doing everything from managing to stomping grapes. One day I was in the basement in my waders, up to my knees in grape pomace—the dried skin and seeds that's left after the wine is floated out of the tank. I was shoveling the pomace out of the tank by hand, when suddenly one of my investors came down the stairs and looked at me. He asked me how it was going and I looked up at him from the tank and told him things were going well, we were profitable, and the crowds seemed to keep coming. He looked down at me and made a suggestion that would spark a whole new mindset for me and my vision for growing that single location into what would eventually expand and be voted as a restaurant industry 'hot concept.' He said to me, 'I don't know if you ever look around here and notice that you are really onto something that could be big. Maybe at some point you should consider climbing out of that tank, putting down the shovel, and getting some other people on board to help take this thing to a whole new level.' So I did, and the rest is history."

A Conversation with Innovator Thomas Stat: On Finding Your Next

Look around at the great innovations of our times ... or any time. They share unique qualities. From the Ford Model T to CNN, FEDEX, and Amazon, to the iPod, Tivo, and Starbucks, these innovations have a lot in common.

- Each emerged as business giants from mere start-ups. None of what was next was developed by the prevailing players.
- Each was far more about the identification, development, and delivery of an opportunity than the solution to an existing problem.
- Each had its origins in deep understanding, inspiration, and insights gained from observing real human behaviors, real context, and real unmet needs. No one could have asked people if they'd like a 24-hour TV news channel, next-day package delivery, or a book from a virtual store. And if they had asked what people wanted, the answers would not likely have been generative or supportive of a next. If the prevailing players *had* asked what people wanted before they offered it to them, these inherent customers would have asked for a CD wallet that could carry fifty discs, a fool-proof VCR, or simply a better cup of coffee. Instead, we experienced innovations that transformed the music industry, disrupted network TV, and created "third place" environments (neither home nor work) where people could relax (with coffee).

(Contnued on next page)

(Contnued from previous page)

- Each addressed impossible questions inspired by extreme synthesis and genomic engineering. What's an alternative to a faster horse? What would happen if you merged the content and immediacy of news radio with the compelling presence of television? In a world of instant faxes and e-mail, could a physical package arrive so quickly as to almost appear telecommuted? What might be the result of marrying the Sears catalog and the Internet? How might we allow people to carry 10,000 CD's around with them and only the songs they actually liked? What if you could put live television on pause as easily as you could put down a book?

  In the end, I believe we are experiencing a remarkable but silent renaissance. It is not an evolution from hunting to agriculture, human capacity to technology, manufacturing to information, or even a revolution from face-to-face encounters to social media. It is a renaissance in approach and intent.

  We have progressed past a time when just "knowing" mattered. We have evolved beyond a reality when "understanding" is the highest value. The new currency of our world is "imagination." From the world of "I know," to "I understand," to "I imagine," we are immersed in a place where innovation outplays invention and opportunities trump solutions. Possibility, feasibility, and the prevailing paradigm are all being questioned. And while theories, methodologies, systems, and programs to implement a good or even great idea abound, the focus is now on creation, conception, and generation. To address such challenges and do this work we need people who can collaborate and synthesize. We need people who, like George Carlin once said, can employ the power of "Vuja De." We need people who can look at something they've seen a hundred times before and, for one instant, see something completely different, imagine the impossible, and imagine the future. We need people who can see beyond a Lamborghini farm tractor and imagine a supercar brand. We need people who can see beyond a black and white cartoon of a funny mouse and imagine an enchanted castle theme park. We need people who can see beyond a wireless communication device and imagine ways to promote health, wellness, and security.

  Imagination and curiosity are in our genes. And genes and elements—the fundamental building blocks of life and matter—are elemental to what can be. They are the building blocks of what *will* be next. At least, that's what I imagine.

Finding Your Next

It's the beginning of a new era. As we guide our organizations toward our next opportunities for competitive advantage and distinction in a world that challenges us to keep up, we can use our business periscopes to take stock of the coming trends and the success of other companies in other industries to make sure we meet, or even anticipate, the needs of the market. The stories of those who have found their next, and have been successful, can inspire us. We can train ourselves to squint at data we see and find patterns that will offer insight into the evolving preferences of the market.

We can accelerate our pace to innovation by grafting ideas from one area of expertise onto another to improve today's business. The Houston Texans did it by offering a better in-seat refreshment option for fans by mastering the art of the McDonald's drive-through. In some cases, we can create a whole new offering, like G.E. did when it grafted its aviation expertise onto its windmills.

Rather than fear the accelerated pace of change, the complexity of the global business environment, or the shift in the customer dynamic characterized by greater transparency and a higher volume on the voice of the crowd, we can embrace those new realities. Our organizations can be trained to use a new set of muscles as we shape the future: we can shift from trying to predict with certainty what will come, and replace that core strength with a new capability, the ability to adapt more rapidly.

Our dashboards should focus on our own six genomic elements: product and service innovation, customer impact, process design, secret sauce, talent and leadership, and trendability.

Most importantly, we can ask ourselves new questions that will leapfrog us forward. Instead of expecting to know with certainty how the current forces will play out, we can organize our corporate mindsets around the question, "Which emerging forces should we be tracking right now to lead our companies toward a thriving future?"

Our companies can become more nimble, more responsive, and more adaptive. Our people can become more outward focused and more alert. And we, as individuals, can listen to our instincts and dive more deeply into our hunches. We can become the kinds of people who can see opportunities as they come our way, the way that Joe Tucci from EMC Corporation saw the future in cloud computing, landing on VMware as an investment, and the way Danae Ringelmann envisioned a new dynamic for funding independent works.

Experience has taught all of us to see new possibilities and recognize new realities, even ones invisible to others. I remember a scene a few years ago, when my friend's four-year-old son was standing at the window and looking out at the neighborhood as the sun was rising. He stared out for a very long time, holding his dad's hand. He finally looked up to his dad and said, "Dad, just what is it about what you're seeing out there that lets you know it's Monday?"

We all recognize the signs and signals that let us understand that it's Monday out there. We can all think differently about tomorrow, knowing that it is our job not to predict what it will be, but that we can adapt to what it might be. We can see the future today in the early signs we observe in other industries right now. We have all invested years of experience mastering the fundamentals of our companies and the competitive forces we face. We have all learned valuable lessons that have given us the talent to now advance to a new set of skills that are finally perfectly suited for this new era.

By recognizing the new realities, shifting our focus to new questions, and tapping into the patterns that are evolving today, we can shape our next.

Bibliography

Chapter 4

2010 Global Automotive Symposium, "China and World Auto Market in the Next Decate," April 22, 2010, http://www.chinaautoreview.com/conference/Introduction.aspx?id=34, accessed January 17, 2011.

Anderson, Chris, "TED Curator Chris Anderson on Crowd Accelerated Innovation," *Wired Magazine*, January 2011, http://www.wired.com/magazine/2010/12/ff_tedvideos/, accessed January 18, 2011.

Apuah, Allan, *Strategic Innovation: New Game Strategies for Competitive Advantage*, New York: Routledge, 2009.

Christensen, Clayton, *The Innovator's Dilemma*, New York: Harper, 2003.

Collins, Glenn, "Run Away to the Circus? No Need, It's Staying Here," *New York Times*, April 28, 2009, www.nytimes.com/2009/04/29/theater/29circ.html, accessed January 22, 2011.

Gaskins, Robert, "PowerPoint at 20: Back to Basics," *Communications of the ACM*, December, 2007.

Holahan, Katherine, "Hanging Out at the E-Mall." *BusinessWeek*. August 9, 2007. http://www.businessweek.com/technology/content/aug2007/tc2007088_912451.htm, accessed January 22, 2011.

"iPad Sales Cross Million Mark Twice as Fast as Original iPhone." *Yahoo News.* May 3, 2010, http://news.yahoo.com/s/ytech_gadg/ 20100503/tc_ytech_gadg/ytech_gadg_tc1901, accessed January 9, 2011.

Johnson, Steven, *Everything Bad is Good for You*, New York: Riverhead Books, 2005.

Kelley, Tom. *The Art of Innovation*, New York: Doubleday, 2001.

Kim, W. Chan, and Renee Mauborgne. *Blue Ocean Strategy*, Cambridge: Harvard University Press, 2005.

Klanten, R., N. Bouquin, S. Ehmann, and F. van Heerden. *Data Flow*, Berlin: Die Gestalten Verlag, 2008.

Kuang, Cliff, "Scott Wilson's iPod Nano Watch Breaks Kickstarter Records, Raises Nearly $1 Million," *Co Design.* December 4, 2010, http://www.fastcodesign.com/1662743/scott-wilsons-nano-watch-breaks-kickstarter-records-raises-almost–300k-in-a-week, accessed January 18, 2011.

"New iPhone Game Bubble Ball Created by 14-Year Old Boy," *myFox-Phoenix.com.* January 21, 2011, http://www.myfoxphoenix.com/ dpps/entertainment/new-iphone-game-bubble-ball-created-by–14-year-old-boy-dpgoh–20110121-fc_11523624, accessed January 22, 2011.

Osterwalder, Alexander, and Yves Pigneur, *Business Model Generation*, New York: Wiley, 2010.

Popa, Bogdan, "Dodge, Mopar Launch iPhone, Android and Blackberry App," *autoevolution.com*, January 20, 2011, http://www.autoevolution.com/news/dodge-mopar-launch-iphone-android-and-blackberry-app–29982.html, accessed January 22, 2011.

Portnoy, Sean, "Does HP Have the Right Strategy to Take on the iPad and Android Tablets?" *ZDNet*, January 15, 2011, http://www.zdnet.com/ blog/computers/does-hp-have-the-right-strategy-to-take-on-the-ipad-and-android-tablets/4852, accessed January 22, 2011.

Ridley, Matt, *The Rational Optimist*, New York: Harper, 2010.

Roam, Dan, *The Back of a Napkin*, New York: Portfolio, 2008.

Scanlon, Jessie, "Coke's New Design Direction," *BusinessWeek*, August 25, 2008, http://www.businessweek.com/innovate/content/aug2008/id20080825_105720.htm, accessed January 18, 2011.

Segal, David, "In Pursuit of the Perfect Brainstorm," *New York Times*, December 16, 2010, http://www.nytimes.com/2010/12/19/magazine/19Industry-t.html?nl5todaysheadlines&emc5a210, accessed January 18, 2011.

Thompson, Clive, "Subscription Artists," *New York Times*, December 13, 2009, http://query.nytimes.com/gst/fullpage.html?res=9F01E6DC1E39F930A25751C1A96F9C8B63, accessed January 18, 2011.

Tufte, Edward, *Envisioning Information*, New York: Graphics Press, 1990.

Ulrich, Karl T., and Steven D. Eppinger, *Product Design and Development*, New York: McGraw Hill, 2007.

Walgreens, "Walgreens and Take Care Clinics Offer Free Blood Glucose and A1C Testing at More Than 1,700 Stores Nationwide as Part of American Diabetes Month," http://news.walgreens.com/article_display.cfm? article_id55357, accessed January 4, 2011.

World Expo Shanghai. "GM Wraps Up Shanghai Expo Presence," World Expo Shanghai, November 1, 2010, http://www.expo2010china.hu, accessed January 17, 2011.

"A Google Guru's Tips for Web Analytics," *Destination CRM*, http://www.destinationcrm.com/Articles/A-Google-Gurus-Tips-for-Web-Analytics–66183.aspx, accessed January 23, 2011.

Chapter 5

Armano, David, "The New Focus Group: The Collective," *BusinessWeek*, January 7, 2009, http://www.businessweek.com/innovate/content/jan2009/id2009017_198183.htm, accessed January 24, 2011.

Beahm, Clyde, interview by Andrea Kates. Interview with Clyde Beahm, former president, Jiffy Lube, January 21, 2011.

Blanchard, Ken, and Sheldon Bowles, *Raving Fans*, New York: William Morrow & Company, 1993.

Bliss World, *Bliss World*, http://www.destinationcrm.com/Articles/A-Google-Gurus-Tips-for-Web-Analytics–66183.aspx, accessed January 23, 2011.

Chafin, Max, "How to Compete in a Reverse Auction," *Inc. Magazine*, May 1, 2007, http://www.inc.com/magazine/20070501/salesmarketing-pricing.html, accessed January 19, 2011.

DreamGrow Social Media, http://www.dreamgrow.com, accessed January 11, 2011.

DreamGrow Social Media, "Review of 22 Social Media and Marketing Trends for 2010," *DreamGrow Social Media*, www.dreamgrow.com/review–22-social-media-marketing-trends-for–2010, accessed January 22, 2011.

Fineberg, Seth, "Time for CRM to Have Its Day," *Accounting Today*, January 24, 2011, http://www.accountingtoday.com/news/Time-CRM-Day–57001–1.html, accessed January 24, 2011.

Flanagan, Chris, "Best Practices in Social Networking," *Business Innovation Factory*, September 15, 2008, http://www.businessinnovationfactory.com/weblog/archives/2008/09/best_practices.html, accessed January 24, 2011.

Ford, Kristin, "Disney Contest: Take Home a DS, See Your Fashion Online If You Win," *OrlandoSentinel*, September 23, 2010, http://thedailydisney.com/blog/2010/09/winners-of-disney-fairies-fashion-design-contest-to-have-outfits-online-take-home-ds-system/, accessed January 11, 2011.

Gaudin, Sharon, "Facebook Dethrones MySpace as Social Network," *ComputerWorld*, November 18, 2010, http://www.computerworld.com/s/article917204/Facebook_dethrones_MySpace_as_social_network?taxonomyid=71, accessed January 11, 2011.

Goebert, Bonnie, and Herma Rosenthal, *Beyond Listening*, New York: John Wiley & Sons, 2002.

Gonzalez, Melissa, "Social Networking in the Corporate World," *Optimum 7*, January 13, 2011, http://wwwoptimum7.com/internet-marketing/social-media/social-networking-in-the-corporate-world.html, accessed January 24, 2011.

Haugen, Dan, "Welcome to Blue Shirt Nation," *TwinCities Business*, April 2009, http://echubasia.com/archives/b2c-b2b-c2c-types-of-e-commerce.html, accessed January 11, 2011.

Hemlock, Doreen, and Karen-Janine Cohen, "Time to Go Global," *South Florida CEO*, May 2007, http://findarticles.com/p/articles/mi_m0OQD/is_5_11/ai_n27242855/, accessed [K1]January 23, 2011.

"Inc. Magazine Fastest Growing Companies List," *Inc. Magazine*, September 1, 2006, http://www.inc.com/magazine/20060901/inc500-methodology.html, accessed January 22, 2011.

Jiffy Lube. *Jiffy Lube Knowledge Center.* http://www.jiffylube.com/education/automotiveeducation.aspx, accessed January 22, 2011.

Kane, Kate A, "Anthropologists Go Native in the Corporate Village," *Fast Company*, October 31, 1996, http://www.fastcompany.com/magazine/05/anthro.html, accessed January 22, 2011.

Kaushik, Avinash, http://www.kaushik.net/avinash/, accessed January 23, 2011.

"Framing the Goal Values Challenge, and Opportunity," *Occam's Razor by Avinash Kaushik*, January 24, 2011, http://www.kaushik.net/avinash/, accessed January 24, 2011.

"NCR Hotel Self-Service Solution Improves Guest Experience and Enhances Efficiencies with Speedy, Simple Registration," *PRLog*, January 20, 2011, http://www.prlog.org/11237410-apa-hotel-deploys-self-service-check-in-kiosks-from-ncr.html, accessed January 21, 2011.

Rae, Jeneanne, "Viewpoint on Innovation," *BusinessWeek*, June 2009, http://www.businessweek.com/innovate/content/jun2009/id20090624_533529.htm, accessed January 23, 2011.

Reichheld, Fred, *The Ultimate Question*. Cambridge: Harvard Business Press, 2006.

Reichheld, Fred, and Rob Markey, "NPS: The Next Six Sigma?" *Business-Week*, September 22, 2006, http://www.businessweek.com/bwdaily/dnflash/content/sep2006/db20060925_265221.htm, accessed January 24, 2011.

Richardson, Lewis Fry, *Weather Prediction by Numerical Processes*, Cambridge: Cambridge University Press, 2007.

Ryan, Eric, *Strategy Is Sexy*, August 7, 2007, http://brandstrategists.blogspot.com/2007/08/eric-ryan-founder-method.html, accessed January 21, 2011.

Scobel, Robert, and Shel Israel, *Naked Conversations*, New York: Wiley, 2006.

Sears. *Sears Chronology*, http://www.searsarchives.com/catalogs/chronology.htm, accessed January 11, 2011.

Spinelli, Stephen, "Make Growth An Obsession," *BusinessWeek*, June 3, 2004, http://www.businessweek.com/smallbiz/content/jun2004/sb2004063_5528.htm, accessed January 24, 2011.

Underhill, Paco, *Why We Buy*, New York: Simon & Schuster, 1999.

Welborn, Mike, interview by Andrea Kates. *President of Global Brand Development, P.F. Chang's China Bistro*, January 2011.

Chapter 6

Arieff, Allison, "Can Airports Be Fun?" *New York Times Opinionator*. December 17, 2010, http://opinionator.blogs.nytimes.com/2010/12/17/can-airports-be-fun/, accessed January 11, 2011.

Baldrige Awards, "History," *Baldridge Awards*, http://www.nist.gov/baldrige/about/history.cfm, accessed January 22, 2011.

Barnes, Brooks, "Disney Tackles Major Theme Park Problem: Lines," *BusinessWeek*, December 27, 2010, http://www.nytimes.com/2010/12/28/business/media/28disney.html?_r=1&ref=disneywaltcompany, accessed January 11, 2011.

Bird, Allyson, "Airport Proposes Face-Lift: Bigger Concourses, Other Improvements Planned; 'Southwest Effect' Cited," *The Post and Courier*, January 22, 2011, http://www.postandcourier.com/news/2011/jan/21/airport-proposes-face-lift/, accessed January 24, 2011.

casasugar. "Cool Idea: Design in a Box," April 8, 2010, http://www.casasugar.com/Interior-Designer-Nicole-Sassamans-Design-Box-Service–8029545, accessed January 11, 2011.

Deming, W. Edwards, *Out of Crisis*, Cambridge: MIT Press, 1982.

Edwards, John, "Siemens Practices What It Produces," *RFID Journal*, June 14, 2010, http://www.rfidjournal.com/article/, accessed January 11, 2011.

Fortune Magazine, "Fortune 500," *Fortune Magazine*, 1965, http://money.cnn.com/magazines/fortune/fortune500_archive/full/1965/, accessed January 22, 2011.

Graham, Ben S, "Paperwork Simplification," *Industrial Management Institute*, May 17, 1950, http://www.worksimp.com/articles/paperwork%20simplification%201950.htm, accessed January 22, 2011.

Hamm, Steve, "To Stay Competitive, Companies are Finding New Ways to Automate Operations, Reuse Technology, and Streamline Processes," *BusinessWeek*, June 21, 2004, http://www.businessweek.com/magazine/content/04_25/b3888614.htm, accessed January 22, 2011.

Hammer, Michael, and James Champy, *Reengineering the Corporation*, New York: Harper Business, 1993.

IBM, "IBM Completes Acquisition of Sterling Commerce." August 27, 2010, http://www–03.ibm.com/press/us/en/pressrelease/32308.wss, accessed January 11, 2011.

Martin, Roger, "Scientific Management Is At Its Peak," *BusinessWeek*, May 21, 2007, http://www.businessweek.com/innovate/content/may2007/id20070521_889911.htm, accessed January 24, 2011.

PRLog, "LRM Interior Design Introduces New Concept: Design in a Box," October 12, 2009, http://www.prlog.org/10373334-lrm-interior-design-introduces-new-concept-design-in-box.html, accessed January 22, 2011.

Shaw, James, *Telecommunications Deregulation*, New York: Artech House, Inc., 1988.

Taylor, Frederick Winslow, *The Principles of Scientific Management*, New York: Harpers Brothers Publishers, 1911.

Tozzi, John, "Borrowing from Avon and Dell to Sell Shirts," *BusinessWeek*, December 9, 2010, http://www.businessweek.com/magazine/content/10_51/b4208026699046.htm, accessed January 22, 2011.

Weiss, Todd R, "United Axes Troubled Baggage System at Denver Airport," *Computerworld*, June 10, 2005, http://www.computerworld. com/s/article/102405/United_axes_troubled_baggage_system_at_Denver_airport, accessed January 11, 2011.

Welch, Jack and Suzy, "The Sig Sigma Shotgun," *BusinessWeek*, September 14, 2007, http://www.businessweek.com/managing/content/sep2007/ca20070914_600442.htm, accessed January 11, 2011.

Chapter 8

Badenhausen, Kurt, "The World's Most Valuable Brands," *Forbes Magazine*, July 28, 2010, http://www.forbes.com/2010/07/28/apple-google-microsoft-ibm-nike-disney-bmw-forbes-cmo-network-most-valuable-brands_print.html, accessed March 30, 2011.

"BP Plc ADR BP Q3 2010 Earnings Call Transcript," *Morningstar.* November 2, 2010, http://www.morningstar.com/1/3/241577-bp-plc-adrbpq3–2010-earnings-call-transcript.html, accessed March 30, 2011.

Craig, Daniel Edward, "Craig's List: Hotel Industry Trends in 2010—Heavenly Deathbeds, Corporate Quilt-making and DickAdvisor," *DanielEdwardCraig*, January 11, 2010, http://www.danieledward-craig.com/2010/01/craigs-list-hotel-industry-trends-in.html (accessed April 1, 2011).

DeLollis, Barbara, "Hotels Roll Out Spruced-Up Deals," *USA Today*, March 21, 2011.

Dover, Mike, and Moffitt, Sean. *Wikibrands*, New York: McGraw-Hill, 2011.

Godin, Seth, *Tribes*, New York: Portfolio, 2008.

James, Andrea, and Richman, Dan. "Seattle P-I to Publish Last Edition," *Seattle Post-Intelligencer*, March 17, 2011, http://www.seattlepi.com/business/403793_piclosure17.html, accessed March 31, 2011.

Lohr, Steve, "G.E. Goes With What It Knows: Making Stuff," *New York Times*, December 4, 2010, http://www.nytimes.com/2010/12/05/business/05ge.html, accessed March 31, 2011.

Ragas, Matthew W., and Bolivar J. Bueno, *The Power of Cult Branding*, New York: Prima Publishing, 2002.

Riley, Charles, "Diet Coke Fizzes Past Pepsi," *CNN Money.com*, March 18, 2011, http://money.cnn.com/2011/03/17/news/companies/coke_pepsi/ index.htm, accessed March 30, 2011.

Schwartz, Matt, "Bargain Junkies Are Beating Retailers At Their Own Game," *Wired Magazine*, November, 2010, http://www.wired.com/magazine/2010/11/ff_hackingretail/, accessed March 30, 2011.

"Twenty Five Things about to go Extinct in North America," *Southern Maryland Online*, November 11, 2008, http://forums.somd.com/life-southern-maryland/160511–25-things-about-go-extinct-america-pt–2-a.html, accessed March 29, 2011.

"Tylenol's Miracle Comeback," *Time Magazine*, October 17, 1983, http://www.time.com/time/magazine/article/0,9171,952212–1,00.html, accessed March 30, 2011.

Winsor, Jon, interview by Andrea Kates, founder, Victors and Spoils, February 16, 2011.

Chapter 9

Anderson, Chris, *The Long Tail: Why the Future of Business is Selling Less of More*, New York: Hyperion, 2006.

Arrington, Michael, "Let's Just Declare TV Dead and Move On," *TechCrunch*, November 27, 2006, http://techcrunch.com/2006/11/27/lets-just-declare-tv-dead-and-move-onhttpwwwtechcrunch-comwp-adminpostphpactioneditpost3865/, accessed February 12, 2011.

Baldrige, "Baldrige Award Winners," 2010, http://www.nist.gov/baldrige/, brand differentiation, accessed February 11, 2011.

Balfour, Frederik, "Analog in a Digital Age: Polaroid Re-launches its Cameras," *Bloomberg Businessweek*, October 13, 2009, http://www.businessweek.com/blogs/eyeonasia/archives/2009/10/analog_in_a_digital_age_polaroid_re-launches_its_cameras.html, accessed February 16, 2011.

Barr, Colin, "CNN.com," *Buffett's $50 Million Credit Card Blunder*, February 27, 2010, http://money.cnn.com/2010/02/27/news/companies/berkshire.geico.fortune/index.htm, accessed March 1, 2010.

Black, Jane. "Come Home To Your Own Chef," *Bloomberg Businessweek*, April 18, 2005, http://www.businessweek.com/magazine/content/05_16/b3929123_mz070.htm, accessed March 1, 2010.

Boutin, Paul, "Venture Beat," *Book Industry Nears Its Napster Moment*, March 1, 2010, http://venturebeat.com/2010/03/01/book-industry-nears-its-napster-moment/, accessed February 12, 2011.

"Brandz Top 100," *Millward Brown*, 2010, http://www.millwardbrown.com/Sites/mbOptimor/Ideas/BrandZ_Rankings/BrandZTop100.aspx, accessed February 11, 2011.

Clark, Dick, "Forecasting in a Challenging Business Environment: Lessons from Procter & Gamble," *Institute of Business Forecasting and Planning*, September 10, 2009, http://bx.businessweek.com/supply-chain-performance/view?url=http%3A%2F%2Fwww.demand-planning.com%2F%3Fp%3D322, accessed February 5, 2011.

Collins, Glenn, "A New Best Friend for the Sommelier," *New York Times*, September 1, 2009, http://www.nytimes.com/2009/09/02/dining/02tside.html, accessed February 16, 2011.

"Fast Company's Most Innovative Companies," *Fast Company*, 2010, http://www.fastcompany.com/mic/2010/industry/most-innovative-consumer-products-companies, accessed February 12, 2011.

Franklin, Paul, "DPreview.com," *Polaroid On the Brink*, July 11, 2011, http://www.dpreview.com/news/0107/01071101polaroid.asp, accessed March 1, 2010.

Gogoi, Pallavi, "Discount Designers," *Bloomberg Businessweek*, September 11, 2008, http://www.businessweek.com/lifestyle/content/sep2008/bw20080911_459145.htm, accessed February 16, 2011.

Herr, Jim, "Five Trends for the Future of Radio," *MediaBeat*, January 12, 2011, http://venturebeat.com/2011/01/12/future-of-radio/ (accessed March 11, 2011).

Ho, Donna, "My It Things," *Norma Kamali Spring 2010 Mercedes Benz Fashion Week New York*, September 18, 2009, http://myitthings.com/FashionWeek/Post/fashion/It-Designer/Norma-Kamali—-Spring–2010—Mercedes-Benz-Fashion-Week-New-York/409182009104668345.htm, accessed March 1, 2010.

"J.D. Power Rankings," *J.D. Power & Associates*, 2011, http://businesscenter.jdpower.com/AboutUs.aspx, accessed February 1, 2011.

Jennings, Lisa, "BJ's Founder to Open Fast-Casual Concept," Nation's Restaurant News, February 16, 2011, http://www.nrn.com/article/bj's-founders-open-fast-casual-concept?ad=news, accessed February 16, 2011.

Keim, Brandon, "Wine Snobs Geek Out on Systems Oenology," *Wired Magazine*, May 25, 2009, http://www.wired.com/wiredscience/2009/05/oenology, accessed February 27, 2010.

Kerr, Jim, "Five Trends for the Future of Radio," *Venture Beat*, January 12, 2011, http://venturebeat.com/2011/01/12/future-of-radio/, accessed February 16, 2011.

Kharif, Olga, "Everyone's Aiming at Satellite Radio," *Bloomberg Businessweek*, January 13, 2006, http://www.businessweek.com/technology/content/jan2006/tc20060113_897980.htm, accessed March 1, 2010.

Macklin, Ben, *Radio Trends: On Air and Online*, Report, New York: eMarketer, 2008.

"Moody's History: A Century of Market Leadership," Moody's, http://www.moodys.com/Pages/atc001.aspx, accessed February 5, 2011.

Mullick, Natasha, "Smartcellar: The Smart Wine List," *Trends Updates*, September 3, 2009, http://trendsupdates.com/smartcellar-the-smart-wine-list/, accessed February 27, 2010.

Nussbaum, Bruce, "Warren Buffett's Bet Against Innovation," *Bloomberg Businessweek*, November 4, 2009, http://www.businessweek.com/innovate/NussbaumOnDesign/archives/2009/11/warren_buffet_i.html, accessed February 16, 2011.

Patterson, Scott, *The Quants: How a New Breed of Math Whizzes Conquered Wall Street and Nearly Destroyed It*, New York: Crown Business, 2010.

Procter & Gamble, "Beginnings of Procter & Gamble," http://www.scienceinthebox.com/en_UK/publications/beginningsofprocter_en.html, accessed February 5, 2011.

"Procter & Gamble," *Ohio History Central*, http://www.ohiohistorycentral.org/entry.php?rec=969, accessed February 6, 2011.

Rae, Jeneanne, "The Importance of Great Customer Experiences," *Bloomberg Businessweek*, November 27, 2006, http://www.businessweek.com/magazine/content/06_48/b4011429.htm, accessed February 16, 2011.

Ranee, Alina, "Questions Raised About School Computer Clubs," *Google News*, April 6, 1987, http://news.google.com/newspapers?id=trkTAAAAIBAJ&sjid=XpADAAAAIBAJ&pg=6935%2C1219321, accessed February 27, 2010.

Riismandel, Paul, "The Decade's Most Important Radio Trends," *Radio Survivor*, January 1, 2010, http://www.radiosurvivor.com/2010/01/01/the-decades-most-important-radio-trends–1-the-birth-and-troubled-childhood-of-satellite-radio/, accessed February 13, 2011.

Rotheli, Tobias F, "Business Forecasting and the Development of Business Cycle Theory," *History of Political Economy*, 39, no. 3 (2007): 481–510.

Sennitt, Andy, "U.S. Digital Satellite Radio Market Forecast to Grow to 55M by 2010," *Radio Netherlands Worldwide*, November 23, 2009, http://blogs.rnw.nl/medianetwork/us-digital-satellite-radio-market-forecast-to-grow-to–55m-by–2010, accessed March 1, 2010.

Svensson, Peter, "Palm's Phone Sales Slump and Its Stock Dives," *Google News*, March 18, 2010, http://www.google.com/hostednews/ap/ article/ALeqM5iqks89HNSdCMXqOQgjb9R3EIL3rAD9EHA2MG6, accessed February 27, 2011.

Thomas, Carolyn, "The Fall of Home Cooking and the Rise of Heart Disease," *My Heart Sisters*, September 22, 2009, http://myheartsisters. org/2009/09/22/fall-of-home-cooking/, accessed March 1, 2010.

"Top Places to Work," *CNN.com*, 2011, http://money.cnn.com/magazines/fortune/bestcompanies/2011/index.html, accessed February 13, 2011.

"Trend Spotting with Author Robyn Waters," *Bloomberg Businessweek*. November 27, 2006. http://www.businessweek.com/magazine/content/06_48/ b4011410.htm (accessed February 16, 2011).

Uston, Ken, "9,250 Apples for the Teacher," *Atari Magazines*, October 1983, http://www.atarimagazines.com/creative/v9n10/178_9250_ Apples_for_the_teac.hp, accessed February 27, 2010.

Webster, Tom, "Podcasting in 2010: The Calm Surface Obscures the Roiling Depths," *Edison Research*, August 13, 2010, http://www.edisonresearch.com/home/archives/2010/08/the_calm_surface_obscu res_the_roiling_depths.php, accessed February 13, 2011.

Worden, Nat, "Netflix Gains As Online Video Wins Fans," *Wall Street Journal*, February 14, 2011, http://online.wsj.com/article/SB100014240 52748703584804576144371093782778.html, accessed February 14, 2011.

Case Study: P.F. Chang's

Buss, Dale, "P.F. Chang's Extends Cool to Supermarkets," *Brand Channel*, November 3, 2010, http://www.brandchannel.com/home/post/2010/11/03/PF-Changs-China-Bistro.aspx, accessed January 31, 2011.

Davis, Lea, "Fast-Casual," *QSR Magazine*, January 16, 2011, http://www.qsrmagazine.com/articles/features/115/fast-casual–1.phtml, accessed January 31, 2011.

Fernandez, Keith, "P.F. Chang's Has Affordable Chinese," *Emirates News*, 26 May, 2010, http://www.emirates247.com/2.291/entertainment/pf-chang-s-has-affordable-chinese–2010–05–31–1.250076, accessed February 3, 2011.

Food Ingredients First, "Unilever Reports 15% Rise in Q4 Profits," February 2, 2011, http://www.foodingredientsfirst.com/news/business/ Unilever-Reports–15-Rise-in-Q4-Profits.html?typeid=5, accessed February 3, 2011.

Fox Restaurant Concepts, *Fox Restaurant Concepts*, http://www.foxrc.com/about_us.html.

Gogoi, Pavali, "Chipotle's IPO One Hot Tamale," *BusinessWeek*, September 23, 2005, http://www.businessweek.com/bwdaily/dnflash/sep2005/nf20050923_1613_db016.htm, accessed January 31, 2011.

Hammonds, Keith H, "Michael Porter's Big Ideas," *Fast Company Magazine*, February 28, 2001.

Horovitz, Bruce, "P.F. Chang's, Burger King, Jamba Juice Sell Frozen Food," *USA Today*, June 9, 2010, http://www.usatoday.com/money/industries/food/2010–06–09-frozenfood09_ST_N.htm, accessed January 31, 2011.

Kogi Barbeque, http://kogibbq.com.

"Panera Company History," *Panera Company*, http://www.panera-bread.com/about/company/mgmt.php, accessed January 31, 2011.

"P.F. Chang's Announces Loan Agreeement with True Food Kitchen," *Business Wire*, August 12, 2009, http://www.bloomberg.com/apps/news?pid=conewsstory&tkr=PFCB:US&sid=aNvXtrEhvBag, accessed January 31, 2011.

"P.F. Chang's Announces International Expansion in Middle East," *Restaurant News Resource*, April 21, 2009, http://www.restaurant-newsresource.com/article38282.html, accessed January 31, 2011.

"P.F. Chang's 3Q Profit Rises on Higher Sales," *National Restaurant News*, October 27, 2010, http://www.nrn.com/article/pf-changs–3q-profit-rises-higher-sales, accessed January 31, 2011.

Porter, Michael, *Competitive Strategy*, New York: The Free Press, 1980.

Scarpa, James, "P.F. Chang's Taneko Japanese Tavern," *Nation's Restaurant News*, February 19, 2007, http://www.nrn.com/article/inside-P.F.-changs-new-taneko-japanese-tavern#ixzz1BhcgV900, accessed January 31, 2011.

Welborn, Mike, interview by Andrea Kates, president of global brand development, P.F. Chang's, January 31, 2011.

COMPANY DESCRIPTION

Want More? Andrea Kates can engage with your organization to bring the principles from her book, *Find Your Next*, to life through her work as a consultant, speaker, and director of CoLabs—hands-on sessions customized to help participants find their next competitive edge. Having led more than 250 strategic projects with major industry leaders, Andrea is a seasoned professional who understands what works. She utilizes a unique mapping tool based on the Business Genome diagnostic survey that will identify your company's DNA and direct them toward cross-industry case studies that offer insightful examples of new strategic directions. To learn more, visit http://www.BusinessGenome.com or email Andrea at akates@businessgenome.com.

Index